"From Elektron to 'e' Commerce"
150 Years of Laying Submarine Cables

Compiled & Edited by Stewart Ash

Picture Editor and Production Manager Gail Clark

Co-Authors Stewart Ash, Mary Godwin, Mick Green, Nigel Morgan and Alcatel's, Communications Department in collaboration with John Waner

Designed & produced by Walker Roast

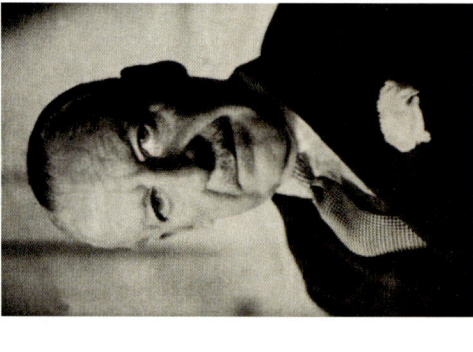

Sir John Pender, GCMG
Chairman of the Telegraph Construction & Maintenance Company - 1864 - 1868, then Chairman of the Eastern Telegraph Company - 1872 - 1896

Sir John Denison-Pender, KCMG, GBE
Managing Director, then Chairman of the Eastern Telegraph Company - 1896 - 1929

Lord John Cuthbert Denison-Pender
1st Baron of Porthcurnow
Managing Director of Imperial and International Communications Ltd, then Chairman of Cable & Wireless - 1929 - 1936

Lord John Jocelyn Denison-Pender
2nd Baron of Porthcurnow
General Manager, then Managing Director of Cable & Wireless - 1936 - 1947

Foreword

It is a real pleasure to write a few words of introduction to a book about the submarine cable industry as four generations of my family made a significant contribution to its development. Our association with telecommunications started with my great great grandfather John Pender. He was already a successful businessman in the Manchester cotton trade when, in June 1852, he invested in, and became a director of, the newly-formed English & Irish Magnetic Telegraph Company. This was no idle speculation, as a shrewd business man, he could see the potential benefits of speedy communication and that preferential access would give him a commercial advantage over his competitors. Through the Magnetic he became aware of the nocturnal tests being carried out by Charles Tilston Bright and his brother Edward. Though not a man of science himself, he grasped the significance of what was being attempted and so when Cyrus Field came to Manchester in 1856 with a proposal to build an Atlantic cable, he was one of the first to invest £1,000 in the company. He played a minor role in the failed 1857 and 1858 attempts, but his appetite had been whetted and his natural tenacity came to the fore. He followed closely the government enquiry into the failures and formed two key opinions. Firstly, that in order to manufacture a successful cable, the whole process should be under the control of a single company and secondly, a ship large enough to lay the entire cable was necessary if they were to succeed. In solving these problems he encountered Daniel Gooch, who had made his fortune with the Great Western Railway and had worked closely with Brunel in building the Great Eastern. The ship was a commercial failure and Gooch was looking for some way to recover his investment. The two joined forces, John Pender formed the manufacturing company and Gooch set up a company to purchase the ship and charter it to the manufacturers. Their first attempt failed with only 600 miles to go. Undaunted, they formed a new company to build a new cable and complete the partially-laid cable. However, investment was difficult to obtain and it was not until John Pender put up a personal guarantee of £250,000 that the money was found. As is well known the 1866 project was a complete success and two cables spanned the Atlantic. Many men would have sat back and reaped the rewards, but this was not John Pender. In 1868, he left the manufacturing side of the business to Daniel Gooch and set out on a grand plan, to build a global network based on submarine cables. By his death in 1896 he had built a network, which was the envy of every other country and would play a leading role in the defence and prosperity of the British Empire.

Much has happened in the ensuing 100 years, international communications has become affordable to everyone and the Internet has revolutionised our way of life. This evolution is all taken for granted by my children and grandchildren, but would not have been possible without the development of submarine cables. We owe a great debt to the vision of those early pioneers and sometimes this is too easily forgotten. The industry founded by John Pender and his contemporaries has a rich and interesting heritage, and it is appropriate that on its 150th anniversary, a book should be produced which gives us the opportunity to pause and reflect on what has gone before.

Lord Pender of Porthcurnow

Pender of Porthcurnow

Lord Pender of Porthcurnow 2000

Contents

Introduction			6
Section 1	*"Realising the Dream"*	Birth of the Industry 1720 - 1856	9
Section 2	*"Throwing a Girdle Round the Earth"*	The Telegraph Era 1856 - 1956	29
Section 3	*"And Nation shall Speak unto Nation"*	The Telephone Era 1956 - 1986	49
Section 4	*"The Light Fantastic"*	The Optical Era 1986 - 2000	69
Acknowledgements			86
Appendix	Industry Chronology 1800 - 2000		87
Bibliography			92
Index			94

Introduction

As we enter the new Millennium, global telecommunications and its importance to international trade is taken for granted. Most of us are comfortable with fax machines, mobile phones, e-mail, conference calls and the Internet. Every day we have more options available to us from our home computer, whether it is to manage our bank accounts and pay our bills, order our groceries, book an air flight or buy a car. Of course, these services did not appear overnight; they took many years to develop.

In the early part of the Industrial Revolution, communication and therefore trade was based on the transportation of written documents by hand, coach and ship. The invention of the electric telegraph changed this dramatically. It is difficult for us to imagine how great an impact this new technology must have had on society and business. However, the quotation, "The world changed forever once information could travel faster than a man on a horse" (attributed to Charles Dickens) indicates the sense of isolation of the times. Not long after the first commercial telegraph services appeared on land an embryo submarine cable industry emerged. It is this industry, now much developed but still very recognizable, that provides the backbone network for the Internet and all other international telecommunications.

The celebrated event which actually launched the submarine cable industry 150 years ago was the laying of a fragile strand of insulated copper under the 22 miles of choppy water that lie between England and France. On 28th August 1850 the fledgling telegraph systems of the two countries were connected, albeit briefly, and that sense of isolation known only to our ancestors had vanished forever.

Whilst it was fashionable for 19th century commentators to marvel at the annihilation of time and distance that the electric telegraph brought about, even the most visionary of them would have been astounded at the content and world access that is now available when the modest home computer is connected to the Internet - a new global economy is being created. But this has all been done before.

In the heady days of the mid-19th century another global economy was in full swing. It was based upon the British Empire, and was fuelled by a revolution in Britain's industry the like of which the world had never seen. Britain in 1850 possessed the two vital ingredients for the new industry of submarine cable to flourish: firstly, it had evolved the necessary technology and manufacturing processes, and secondly, it possessed a globally-based economy poised to make good use of all the benefits that this new communication medium would bring. The characters involved were not slow to capitalise upon their fortunate situation and, as a result, for about a century the global development of submarine cables was largely a British story.

It is therefore appropriate that four of today's major companies, all of whom can trace their histories back to the infancy of this industry, should come together to tell of their heritage and of those whose vision, courage and creativity launched their common enterprise. These four companies, which still operate from Britain as world-leaders in the global submarine cable business, are:

Alcatel

On the manufacturing side, the history of Alcatel's submarine networks activities go back to the original Gutta Percha Company, which in 1850, made the Brett brothers' first submarine cable connecting England and France, and to Küper & Company, a small manufacturer of wire ropes for collieries which armoured some of the replacement cable of 1851. Alcatel manufactures world-wide, having facilities in America, Australia and France in addition to its plant at Greenwich, which stands on the same site as the Telegraph Construction & Maintenance Company factory, which dominated the business for almost a century.

British Telecom (BT)

British Telecom traces its history back to John and Jacob Brett, who laid the first (short-lived) England to France cable of 1850 and, via the nationalisation of the UK's Telegraph companies in 1868, to the Submarine Telegraph Company of T R Crampton and the Brett brothers, who laid its successful replacement in 1851.

It goes without saying that many costly lessons would have to be learned by the new industry before it would reach the sophistication of the optical cables available at the start of this Millennium. Over its 150-year history, the demand for ever more capacity has rarely given the cable manufacturers much respite; the arcane art of cable-laying and repairing has seen its mysteries yield to science and technology; and the reintroduction of competition has wrought a total change in the service offered to the public.

Our story divides neatly into four periods. The first section entitled "Birth of the Industry" covers the early experiments and developments, which take us up to 1856, just prior to the first

Cable & Wireless (C&W)

Cable & Wireless traces its antecedents back to Sir John Pender, a wealthy cotton merchant who became a Director and shareholder in two of the earliest telegraph companies: the English and Irish Magnetic Telegraph Company in 1852 and the Atlantic Telegraph Company in 1856. He later founded the companies that would eventually become the nucleus of the C&W group.

Global Marine Systems Limited (Global Marine)

Global Marine Systems Limited, originally Cable & Wireless (Marine), was first formed in 1983 by reorganisation within its parent company. It remained a wholly owned subsidiary of Cable & Wireless until 1999 when it was acquired by the Bermuda-based Global Crossing Ltd.

Atlantic telegraph cable. The second section, the "Telegraph Era", covers the development of the global network of telegraph cables, a 100-year period which effectively came to a close with the first Atlantic telephone cable TAT-1 in 1956. Section three, the "Telephone Era", covers the 30-year period of telephone cables which came to an abrupt end in 1986 with the inception of fibre optic submarine cables. Finally, section four, the "Optical Era", covers the amazing advances in technology that have led to the modern high capacity submarine network on which the global economy now relies so heavily.

Who would dare speculate where the submarine cable industry will find itself in a further 150 years?

7

"Realising the Dream"

Section 1 Birth of the Industry 1720 - 1856

The Possibility of Electric Telegraphy

We all know that a magnet will attract iron and a rubbed glass rod will attract a feather, but no one knows at what date the phenomenon of "attraction" was first discovered.

The subject remained a mystery until 1600. In that year William Gilbert, physician to Queen Elizabeth, published his *De Magnete* in which he described most of the known phenomena to do with attraction and distinguished between magnetic attraction (to do with lodestones and iron) and the attraction associated with amber which he called after its Greek name *elektron*, thereby giving us the term "electricity".

Gilbert's work sparked off a lot of interest, but little actual progress was made until 1720 when Stephen Gray, a London pensioner, formulated for the first time the principles of electric conduction and insulation. This was quickly followed by Benjamin Franklin in Philadelphia who discovered that electricity consisted of a single "fluid" with "positive" and "negative" aspects. He also worked out the explanation for the Leyden Jar whose ability to store electricity was a mystery.

The first written proposal to suggest employing electricity for the transmission of intelligence appeared in 1753 as a letter to the *Scots' Magazine*, signed CM. There were other similar proposals, all based on frictional or what today we would call "static" electricity. As late as 1816 Francis Ronalds was successfully demonstrating an electrostatic telegraph in his back garden and experimenting on the velocity of electricity, though by then it was becoming clear that static electricity was unsuitable for public telegraphy. Ronalds also published a detailed proposal for a

Left
Dover/Calais Cable lay from the Blazer in 1851

Above
Francis Ronalds' eight miles of wire set up on his lawn for experiments on the velocity of electricity

system of "... electrical conversazione offices communicating with each other all over the kingdom ..." which was the first published work to set down the principles for building and operating a national telegraph network.

By 1800, Alessandro Volta had developed the so-called Voltaic Pile, which was the direct forerunner of today's electric battery. It was subsequently improved by Professor Daniell of King's College, and as a result of the work of Humphrey Davey and Michael Faraday at the Royal Institution, some of the effects of the new electricity, particularly its chemical effects, came to be understood quite rapidly. The Voltaic Pile, which provided a relatively constant source of electric current, seemed to have an advantage over electrostatic generators and was responsible for various telegraphic experiments using frogs legs and its ability to decompose water. But for public telegraphy an easier means of detection was needed.

Although a link between electricity and magnetism had long been suspected, it was not until 1820 that Christian Oersted at the University of Copenhagen announced his discovery of "electromagnetism". He had noticed that a wire in which an electric current was flowing was able to move a compass needle. This discovery provided not only the missing link between electricity and magnetism, but also a simple device which could be used for indicating telegraphic signals. It seems that Oersted did not fully understand his observations and it was left to the French mathematician André-Marie Ampère to provide the theoretical explanations. He also proposed a number of electromagnetic instruments, notably the spiral or helix of wire which would form a vital component in every electric telegraph system.

Later during 1820, by applying Ampère's laws, a chemist in Halle, Johann Schweigger, realised that by increasing the number of turns in his spiral or "coil" the sensitivity of the simple compass needle instrument could be greatly increased. Schweigger's "Multiplier" quickly evolved into the ubiquitous "galvanometer" and is particularly significant as the forerunner of the famous mirror galvanometer which William Thomson (later Lord Kelvin) developed less than 40 years later for long oceanic cables.

By the close of 1820, all the essential components for a basic electric telegraph system had been discovered: the battery would provide a continuous electric current; insulated wires would transmit the current to a remote point; and it could be detected easily via its magnetic effect on a needle.

Cooke & Wheatstone's Electric Telegraph

In 1812 Baron Pawel Lwowitsch Schilling von Canstatt, an attaché of the Russian Embassy in Munich, had succeeded in exploding powder mines with an insulated wire laid across the River Neva, near St Petersburg. By about the mid-1820s he had moved on to the electric telegraph.

Schilling's telegraph used Schweigger's Multiplier to move a magnetic needle to which was attached a little paper disc, black on one side and white on the other, so that, with several discs, combinations of black and white would represent letters of the alphabet and numbers. It was a clever system. In fact, Professor Müncke of Heidelberg University was so impressed that he had a copy of Schilling's apparatus constructed for his lectures.

As the 1830s dawned, new telegraphic inventions started to appear in profusion. In particular, Edward Davey, a chemist, probably came closest to realising a complete, practical telegraph system but unfortunately he had to leave the country due to marital problems, opening the way for his main competitors who were Cooke and Wheatstone. On marine telegraphs, Davey had said in his opposition to Cooke and Wheatstone's 1837 patent application:

Birth of the Industry 1720 - 1856

Communications may be effected through, or under, the water by enclosing the conductors in ropes well coated, or soaked, in an insulating and protecting varnish, such as melted caoutchouc..... we may have an air-tight and water-tight electrical renewing apparatus at each requisite interval.

So, Davey not only had the insulated submarine cable clearly in mind, but he also envisaged relay-type "repeaters" to boost the weak signals before sending them on their way into the next cable section.

In 1836 one of Professor Müncke's physics lectures was attended by an Englishman, William Fothergill Cooke, an army man who was in Heidelberg to study for a new career in anatomical modelling. Cooke saw Müncke's replica of Schilling's telegraph in operation and immediately embarked on a further change of direction for which he possessed no training at all. Cooke, in reality a businessman, recognised an opportunity when he saw one.

Cooke built a number of instruments but, not surprisingly, they did not work very well. Even a consultation with Michael Faraday did not really help him. Nevertheless, with little to sell, he was able to persuade the directors of the Liverpool and Manchester Railway to allow him to experiment on a mile-long tunnel in Liverpool which required a communication system between the stationary engine driver at one end and Lime Street station at the other. Desperate to make his system work, he called upon most of London's scientific community until finally coming across "an extraordinary fellow" called Charles Wheatstone, Professor of Experimental Philosophy at King's College, London. Wheatstone was an established figure

Left
William Fothergill Cooke

Below left
Charles Wheatstone

Right
Powered by the newly-invented steam locomotive, the railways spread rapidly across the country from 1830. About a decade later they provided routes for the electrical telegraph

in the scientific world and already deeply involved with electric telegraphy.

From the start their relationship was an uneasy one. The businessman and the scientist needed each other, but neither was prepared to recognise the fact. Nevertheless, in March 1837 they agreed to form a partnership. They applied for a joint patent in May which, after Edward Davey's unsuccessful challenge, was sealed on 12th June that year. It was the world's first patent for an electric telegraph. Significantly, when the specification came to be filed some months later, all the instruments cited were Wheatstone's.

Cooke laid on some major demonstrations for railway companies including one between Euston and Camden Town in the autumn of 1837 and another along Brunel's Great Western Railway line from Paddington to West Drayton in May 1838. By 1842 he had persuaded the directors to allow him to extend the system to Slough and offered to carry railway messages free if the railway would lease him the route for a nominal charge. Agreement was reached, and in 1843 Cooke opened his own telegraph service to the public, charging one shilling per message irrespective of length.

The following year the Admiralty signed up with Cooke for a telegraph between London and Portsmouth and in 1845 Cooke's telegraph on the Great Western Railway helped to apprehend the murderer John Tawell to great public astonishment.

In 1840 the London and Blackwall Railway had opened with Cooke's electric telegraph to control its operations, followed soon after by the Yarmouth & Norwich Railway. Gradually the electric telegraph became an indispensable part of every railway system.

Wheatstone's contribution to the partnership was to design instruments and telegraph circuits in the firm knowledge that they would work properly, whilst Cooke supplied the business skills. Despite their personal differences, the Cooke and Wheatstone telegraph system was a commercial success and by 1845 its expansion required the formation of a company - the Electric Telegraph Company (later known as the "Electric") - which Cooke formed with two "influential capitalists", the MP and financier John Lewis Ricardo and his friend George Parker Bidder. The Electric was incorporated during 1846 and grew into the United Kingdom's largest telegraph company before becoming part of the Post Office in 1870 when all the country's telegraph companies were finally nationalised.

Wheatstone sold specified parts of his patents to the new company, his partnership with Cooke came to an end, and he continued to develop his many other scientific interests at King's College. One of these was submarine telegraph cables.

Gutta Percha, Bottle Stoppers and the Problem of Insulation

In 1840 and 1841 Wheatstone visited Brussels and Paris to discuss a cross-Channel submarine telegraph cable. His drawings for the project are impressive. The cable was to contain seven conductors, insulated with yarn saturated in boiled tar and protected by iron wire. He showed the cable design, the cable-making machine, a profile of the seabed, depth soundings between Dover and Cap Gris-Nez, cable-laying

cables in Swansea Bay and in 1845 Ezra Cornell laid a twelve-mile cable in the relatively sheltered waters of the River Hudson to connect Fort Lee and New York which worked well until broken by ice the following year.

In the race to lay a cable across the Channel, Wheatstone had a competitor - Charles West - who held a patent for the insulation of wires with India rubber. He conducted trials in Portsmouth harbour of a cable insulated with India rubber made by Messrs S W Silver & Co. They were originally garment makers in Greenwich, who from 1844 began making waterproof clothing, vulcanised products and rubber-covered electrical conductors. The trials worked well, but the project fell through when the Electric declined to support it. By 1845 Wheatstone was thinking of insulating his Channel cable with a new material from Malaya called Gutta Percha that had recently been discussed at the Society of Arts, but he could not obtain enough of it and let his plans lapse.

Gutta Percha is the dried sap from an immense tree, the *Isonandra Gutta*, which occurs in many parts of South-East Asia. It is closely related to caoutchouc or rubber with the unusual property that it can be shaped by softening in hot water, hardening again when cool. It seems to have been brought to the West by at least three people round about 1843. One of these, Dr William Montgomerie, a surgeon working for the East India Company, sent samples to the Society of Arts, pointing out its properties and possible uses. The following year he brought back more samples, one of which found its way to an artist and inventor called Charles Hancock who incorporated the new material into a patent specification for bottle stoppers. Not long afterwards, Hancock was approached by a Dublin chemist,

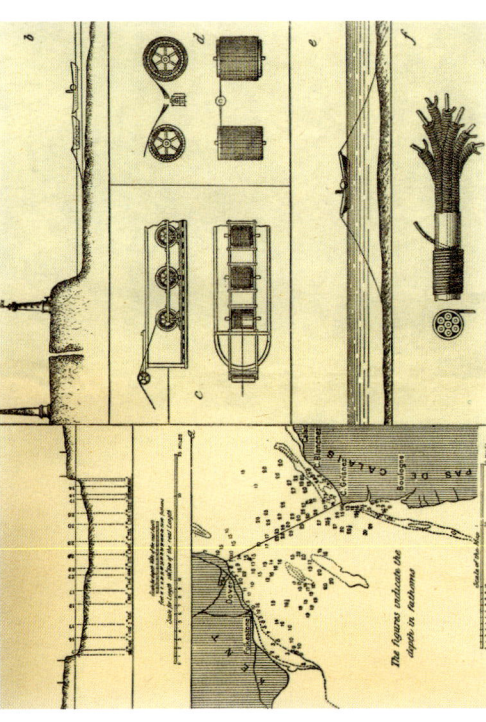

Above
One of Wheatstone's drawings for a cross-Channel submarine telegraph cable

machinery and its installation on a barge. It is clear that he had given it all a lot of thought. When in 1843 Wheatstone assigned most of his patent rights to Cooke in exchange for royalties, he took care that the assignment did not prevent him from "establishing electric telegraph communication between the coasts of England and France ... for his own exclusive profit".

In 1843, Samuel Morse was able to send electric currents through a cable insulated with hemp soaked in tar and pitch surrounded by a layer of India rubber across New York harbour. In 1844, Wheatstone conducted experiments with submarine

Birth of the Industry 1720 - 1856

Henry Bewley, who was experimenting with bottling fizzy water and needed a good stopper. So on 4th February 1845, Hancock and Bewley agreed to work Hancock's patent for their joint benefit. This date fixes the birth of the Gutta Percha Company which later became the heart of an industry supplying the world with submarine cables.

Wheatstone had probably heard the original suggestion to use Gutta Percha for insulating cables which came from Michael Faraday to William Siemens. Siemens had arrived in London in 1843 to represent the Berlin firm of Siemens & Halske. They quickly adopted the idea and were soon installing great lengths of insulated telegraph wire underground in Germany and Prussia.

The Gutta Percha Company prospered and, in addition to insulated telegraph wire, supplied Victorian society with a tremendous range of domestic products such as picture frames, soles for shoes and conversation tubes for noisy railway journeys. After a while Hancock and Bewley fell out over a patent dispute, and since Bewley's backers were providing most of the funds, Charles and his brother Walter were paid off and left.

In 1848 the Gutta Percha Company received its first order for a submarine cable (a two-mile length) from C V Walker, a Fellow of the Royal Society and telegraphic engineer to the South Eastern Railway Company. He had used Gutta Percha insulation on telegraphic wires through all the wet railway tunnels for which he was responsible and now wanted to experiment in the Channel.

Above
The collection of Gutta Percha from the Isonandra tree

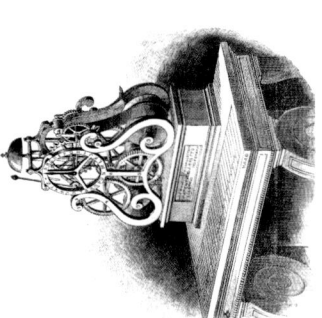

Above left
John Watkins Brett

Above
Jacob Brett

Left
Jacob Brett's Electric Telegraph

On 10th January 1849, using a small vessel, the *Princess Clementine*, Walker laid the cable from the beach at Folkestone out in a loop. One end was connected via a telegraph line to London whilst Walker remained on board the *Princess Clementine* where he was able to exchange messages a total distance of 85 miles, of which two miles were under water.

Gutta Percha proved to be an ideal insulator and for about a century would remain the prime material for insulating submarine cables. India rubber was also used, but never as widely.

The Founding of an Industry

By the late 1840s, the basic technology existed, albeit in primitive form, to make submarine cables. It was not long before promoters came forward with the first commercial submarine cable project - to connect England with France. They were two enthusiastic brothers Jacob and John Watkins Brett.

On 16th June 1845 Jacob Brett registered the General Oceanic Telegraphic Company but the registration lapsed due to government inertia. Later that year he patented a version of a weight-driven printing electric telegraph instrument invented by Professor Royal E House of America which he proposed to modify to work on submarine cables. Jacob was then joined by his brother John and in 1846 they registered the General Oceanic & Subterranean Electric Printing Telegraph Company:

To establish a telegraphic communication from the British Islands across the Atlantic Ocean to Nova Scotia and Canada and establishing electric communication by land and sea with the Colonies.

The landing of a submarine cable in England only required the formality of a licence, which did not exclude others from obtaining licences of their own. In France, however, a concession, once obtained, granted exclusive rights for ten years. The Brett brothers obtained their first French concession in 1847 after a lengthy period of negotiation but let it lapse. They were able to renew it again in August 1849 for a further ten-year period with the proviso that it too would lapse if communication had not been established by 1st September 1850.

So the brothers had a year to organise themselves, get the cable manufactured and lay it. They formed a third company, the English Channel Submarine Telegraph Company in which four shareholders put up £500 each - John Watkins Brett, Charles Fox, Francis Edwards and Charlton James Wollaston, the latter acting as engineer.

They placed an order with the Gutta Percha Company for 25 nautical miles of copper wire covered with Gutta Percha to make a "cable" with an overall diameter of half an inch. This, without any other protection, would be laid directly in the sea. Needless to say, the primitive manufacturing processes and the many joints provided plenty of scope for faults, and it is remarkable that it did in fact work.

The plan for laying the cable was to use a small paddle-driven steam tug, the *Goliath*, chartered for the purpose. The *Goliath* came up the Thames and individual coils of cable were jointed and wound on to a huge horizontal drum seven feet in diameter and 15 feet in length mounted across the vessel just behind the funnel. At Dover and at Calais, special shore-end cables were installed before the *Goliath*'s arrival.

The weather on 28th August 1850 was fine as the *Goliath* steamed out of Dover harbour escorted by HM Surveying Paddleship *Widgeon*. Having picked up and jointed the main cable onto the shore end, the little flotilla steamed off for France with the drum rotating steadily and the fragile cable streaming out behind. It was known that a cable of this design would be almost buoyant in water, so lead weights were attached to sink it. Gradually the French coast drew nearer, and at 6.00 pm the same day, the *Goliath* dropped anchor near the buoy marking the shore-end at Cap Gris-Nez just outside Calais.

The modified House printing telegraph instrument was hooked up to the cable end for tests back to England. The House machine would normally print on to a paper tape at about fifteen words per minute, but on this occasion little of what emerged was intelligible although signals of a kind were present. With daylight disappearing the "final splice" was made and the cable consigned to the sea. It was very disappointing.

Once the cable party was ashore in France, the more sensitive Cooke and Wheatstone needle telegraph instrument was connected up and the first signals exchanged with England. Later that evening it appears the House machine was persuaded to work, perhaps more slowly, and various messages were sent and received, one of which was from John Brett to the soon-to-be Emperor, Napoleon III, who had been taking a keen interest in the project. That day, 28th August 1850, 150 years ago, marks the start of the submarine cable industry in Britain.

The next morning *The Times* reported: "The electric telegraph appears to us more like a miracle than any scientific discovery or mechanical achievement of our time", but unfortunately by

*Left
Goliath laying the first
cross-Channel cable
escorted by HM Surveying
Paddleship Widgeon.*

Birth of the Industry 1720 - 1856

the following morning the cable was found not to be working. The return to Boulogne of a fisherman bearing a considerable length of what he thought was a new variety of seaweed with a gold centre, appeared to explain the failure although, more prosaically, when the cable was raised for inspection a few days later, the lead tube that formed the protection for the shore end was found to have been badly damaged on rocks, about two hundred yards from the beach.

So, short-lived though it was, the Brett brothers' first cable had demonstrated conclusively that telegraphic communication could be made to work under water over quite long distances. But it was found that, because the cable was acting as a store for electricity (like a "capacitor"), the transmitted signals became delayed and distorted, and this was probably the reason the House instrument had not worked well initially. There was much still to learn.

John Brett managed to get the French government to roll his concession forward for another ten-year period provided a cable connecting the two countries was working by 1st October 1851. With so recent a failure, raising the money proved difficult. Eventually, just seven weeks before the French deadline, the famous railway engineer Thomas Russell Crampton personally put down half the £15,000 required and raised the balance between Lord de Mauley and Sir James Carmichael, the three of them forming the board of a new company, the Submarine Telegraph Company.

Thus the manufacture of the cable could now proceed. Crampton made it his business to see that the lessons of the Bretts' first cable were thoroughly learnt. Whilst a cable hardly more substantial than domestic lighting cable might function well electrically, it certainly needed protecting in the relatively shallow waters between England and France. The new cable would be constructed as a Gutta Percha "core" similar to the 1850 cable, but with a protective sheath of iron wires.

The core of the new cable in fact contained four copper wires, each individually covered with two layers of Gutta Percha applied with Hancock's new extrusion process (similar to that of Siemens & Halske) and laid up together. This was then taken to the nearby firm of Wilkins and Weatherley in Wapping, who made wire ropes for collieries. They began applying a helical layer of ten galvanised iron wires to produce a well-armoured cable with an overall diameter of about 1.3 inches.

Not long afterwards an injunction arrived from another firm of wire-rope makers, Messrs Newall & Company of Gateshead claiming patent infringement. Eventually it was agreed that Newall's staff would complete the contract at Wapping. In due course the cable was manhandled out of the factory into a hulk called *Blazer* which was then towed to the lighthouse at South Foreland for cable-laying operations to begin. It was 25th September 1851. This time the weather was not good, the tugs drifted, cable was wasted and the last mile had to be temporarily completed using four Gutta Percha coated wires. Nevertheless, Crampton was able to tell scientists present at the closing ceremony of the Great Exhibition, in Hyde Park, that the cable had been successfully laid to France.

On 19th October 1851, the temporary wires were replaced with a further mile of armoured cable using a steam tug called *Red Rover*. Although technically the 1st October deadline had been

Right
Cable making machine probably at Wilkins & Weatherley's wire rope factory, Wapping, 1851

exceeded, the French government fulfilled its side of the bargain and granted the ten-year concession.

The Crampton-Brett cable continued in service for many years, eventually finding its way, much repaired, into the Post Office in 1890 when the assets of the Submarine Telegraph Company were taken over as part of the nationalisation of the UK's telegraph system, begun in 1868. During its life the Submarine Telegraph Company laid a total of thirteen cables, mostly to continental Europe and in 1894 its cableship, the *Lady Carmichael*, became the first in the Post Office's line of cableships bearing the name *Alert*.

The industry in the 1850s consisted of a number of cable makers, to whom the Gutta Percha Company was the sole supplier of Gutta Percha insulated core. Some of the cable makers were originally manufacturers of wire rope (mostly used in coal-mines) and others emerged from the rapidly developing telegraph industry itself.

The finished cable of the 1850s was still a far from perfect product. The basic copper wire was described as "hard, brittle, soft and rotten" and the Gutta Percha covering had bumps and seams in it. The wire-rope companies suffered from the breakage of armour wires which also frequently damaged the core itself from lumps of galvanising as well as from knots and hard bits in its own yarn covering. The factories themselves were dirty, noisy and badly laid out, and possessed only the crudest arrangements for handling and storing delicate core and finished cable.

Gradually the industry went through a process of rationalisation, although it was not until the disaster of the first transatlantic cable in 1857 and 1858 that the paramount importance of quality was fully appreciated. The quality problem was finally solved in 1864 in preparation for a second transatlantic cable when the Gutta Percha Company and Glass, Elliot & Company were brought together under the responsibility of a single management as the Telegraph Construction & Maintenance Company, sometimes known as "TC&M" or "Telcon".

Towards an Atlantic Cable

Although many failures lay in store for hopeful cable "projectors", the highly successful Crampton-Brett cable of 1851 acted as an ongoing inspiration to all of them. But over how great a distance could electrical communication be persuaded to work? Many, including the Brett brothers, had the Atlantic in mind.

In 1847 a young man, Charles Tilston Bright, took a job with the Electric (William Fothergill Cooke's Electric Telegraph Company) when he was still only 15. His brother Edward, who also joined the company shortly afterwards, was 16. Charles' first job was as a telegraph operator working in a signal box at Harrow station on the London and North-Western Railway which had engaged the Electric to supply telegraphic services under contract. Within a year, he and Edward had become inventors and applied for a patent, which was finally granted on 21st October 1852.

The patent contained 24 individual inventions, the most significant of which was a system for testing telegraph lines to localise faults from a distance. Without the ability to make such "localisation tests", investors would not have had the confidence to advance the enormous sums of money that "wiring the world" with submarine telegraph cables would require.

The other significant event of 1852 for Charles was the invitation to become Chief Engineer of the English and Irish Magnetic Telegraph Company (later known as the "Magnetic") - an extraordinary appointment for a 20 year old. His brother Edward was already the company's manager. The Magnetic was strongly supported by northern capital, one of its investors being a wealthy Scottish cotton merchant who had moved to Manchester called John (later Sir John) Pender. By the time of his death in 1896, Pender's companies controlled one third of the world's submarine telegraph system and formed the nucleus of what was later to become Cable & Wireless.

Pitted against the Electric with its head start of six years, the Magnetic had to promote itself energetically and a bit differently. Its head office was in Liverpool, not London, its electric signals were generated by magneto-electric induction, and its main lines in England were largely laid underground. Charles Tilston Bright extended the Magnetic's system on a vast scale throughout the UK and (awaiting a submarine connection) throughout Ireland.

Following their success with the 1851 Channel cable, the submarine cable making side of Newall's business developed quickly. In 1853, it made a heavy cable to connect the

Above
Charles Tilston Bright

Magnetic's English and Irish telegraph systems across the 23 miles of the Irish Channel between Portpatrick and Donaghadee. This was the fourth attempt at connecting England with Ireland and only the third successful submarine cable to have been laid - an outcome justifiably attributed in large measure to Charles Tilston Bright's direct supervision.

It will be recalled that when Professor Royal E House's modified printing telegraph instrument was connected by the Brett brothers in 1850 to their first cable to France, little of what had emerged from the machine was intelligible. Yet there had been signals present which could be read more slowly with a needle instrument. This was the mystery to which Charles and Edward Bright now turned their attention.

Transmission speed was vital to a telegraph company. With Charles Tilston Bright's cable to Ireland, the correct conditions became available for the first time to make the necessary investigations for the Atlantic cable, which the two brothers were planning. They discovered that a cable laid underground behaves, electrically speaking, very much like a cable laid in the sea. With the English and Irish systems connected together, the Bright brothers possessed in electrical terms the equivalent of thousands of miles of submarine cable.

From 1853 to 1855, while the Bright brothers were advancing the Magnetic's telegraph system to the west of Ireland to be ready to interconnect with a transatlantic cable, they were also conducting transmission experiments. These experiments necessitated a re-arrangement in the way the underground cables were connected. For example, instead of 10 circuits from London to Manchester of roughly 200 miles, they might reconnect the wires in series so as to provide one long circuit of 2,000 miles. At other times long circuits were arranged by looping back and forth in the various underground cables (and their new 1853 submarine cable) between London and Dublin. In this way a subterranean circuit of Atlantic proportions could be represented. Since all this work had to be undertaken at night or on Sundays to avoid interrupting normal traffic on the network, the Bright brothers suffered many nights with little or no sleep.

Following Wheatstone's work in the mid-1830s, everyone was just getting used to the idea that electricity travelled along a wire (in air) inconceivably quickly - in fact at a speed comparable with the velocity of light. But it certainly didn't seem to be going at anything like that speed in a subterranean or a submarine cable, and it was years before telegraph engineers fully understood the problem. The first paper on this subject was presented by Edward Bright to a meeting in Liverpool of the British Association for the Advancement of Science in 1854. The following year Charles and Edward Bright took out a patent for the special methods they had developed for signalling through long underground and undersea conductors - methods which were certainly going to be needed for an Atlantic cable.

In 1855, Edward Bright accompanied by some of the staff from the Magnetic, chartered a small fishing smack to survey the south-west coast of Ireland for a suitable landing for an Atlantic cable and in due course reported to his brother that Valentia was the most suitable landing point. In fact, it was almost exactly the nearest point to the American continent in Newfoundland.

It was a similar story on the other side of the Atlantic. The first task was to provide a reliable telegraph connection between New York and St John's, Newfoundland, the nearest point where an Atlantic cable could land. This was no easy project; Newfoundland had no roads or railways and the country was mostly impenetrable forest, remote and exceedingly uninviting. In fact the difficulties had already overcome an English engineer and entrepreneur, Fredrick Gisborne, who had been attempting to bring St John's onto the telegraph network. His plan was to exchange messages with passing ships using water-tight barrels and save the additional two days' passage to Halifax where they would normally be dropped off or collected.

Gisborne had surveyed the land route across Newfoundland, obtained a concession from the legislature and visited England to purchase a submarine cable to link Prince Edward Island to New Brunswick, which was laid in late 1852. It was the first commercial submarine cable to be installed on the American continent. But the difficulties in constructing the land-line were immense, and when his backers withdrew he was forced into bankruptcy. Barely 40 miles out of 400 had been completed. Almost destitute, he went to New York, hoping that something would turn up. It did. A chance meeting brought him to the notice of a wealthy paper merchant called Cyrus W Field who, at the age of thirty-five, was already retired. Field was initially a bit cool about Gisborne's project but his imagination was fired later that same evening, when he was casually scanning the globe in his palatial library and realised the potentially enormous significance of linking the telegraph systems of the two continents with a cable across the Atlantic Ocean. This would be a great deal better than trying to catch message barrels bobbing around in the open sea - but what a vision!

Field realised that such a project, if possible at all, would be an immense undertaking, but it was just as well that he had no idea of the thirteen years of unremitting toil, heartache and expense that awaited him. For now, this was just the kind of project a retired man of wealth was looking for. He was hooked, but cautious. The next day he wrote to the foremost experts in America - to Samuel F B Morse regarding the electrical feasibility of the idea, and to Lieutenant Matthew F Maury regarding its marine aspects. He was soon in possession of two favourable replies. It turned out that Morse, the architect of the American telegraph system, having experimented with submarine cables had been predicting an Atlantic telegraph for over ten years. More surprisingly, it appeared that the National Observatory under Maury's direction (and with immaculate timing) had just completed a line of soundings and core samples from Newfoundland to the south-west coast of Ireland using US Navy ships. In fact, he had even named part of this route the "Telegraph Plateau". It must have seemed to Field that this was a project destined to happen.

So, armed with two undeniably respectable opinions, Field was able to gather the support of a number of capitalist friends and form the "New York, Newfoundland and London Telegraph Company". What incredible significance was bound up in those two little words "and London"! The company commenced by taking over Gisborne's bankrupt operation and discharging his debts.

Field now came face to face with reality as working in Newfoundland quickly turned into a nightmare. As anticipated, the worst bit was the 400 miles of road and telegraph line that had to be constructed between St John's and Cape Ray, and,

Birth of the Industry 1720 - 1856

and New York. Part of the system was 60 miles of submarine cable connecting Cape Ray at the south-west extremity of Newfoundland with the island of Cape Breton across the Gulf of St Lawrence. It was the ordering of this cable that, in 1854, brought Cyrus Field on the first of many visits to England. He placed the order with Glass, Elliot & Company of Greenwich, a newly established firm specialising in making submarine cables, formed out of the wire-rope business of Küper & Company.

While in England, Field made two key contacts. First he met John Watkins Brett of the Submarine Telegraph Company and pioneer of the first and second England to France cables. He was currently engaged on laying cables in the Mediterranean and had also been planning to construct an Atlantic cable, instead of which he invested £5,000 in Field's company. Field's other meeting whilst on that first visit to England was with Charles Tilston Bright of the Magnetic. The Magnetic's growth in just a few short years plus all its experimental work for an Atlantic cable would have made a deep impression on him. So Cyrus Field returned to New York not only having ordered a cable, but also bearing the news that Britain was well advanced in developing the technology for an Atlantic cable and that capitalists and their telegraph companies were pressing forward rapidly.

The cable Field had ordered was delivered the following year. Unfortunately due to bad weather and inexperience it was partially lost in the attempt to lay it across the Gulf of St Lawrence, but in 1856 Glass, Elliot & Company made and laid a replacement cable which lasted for over ten years. By the time the Gulf of St Lawrence had been bridged the land lines were also in place and telegrams could be sent as far as St John's.

Above
Cyrus W Field

nearly as bad, was a similar link 140 miles long on Cape Breton Island. Eventually after two years and a cost of over US$ 2,000,000, they had completed Gisborne's original project and telegrams could be sent the 2,000 miles between St John's

Above: SS Persian loading John Brett's Mediterranean cable at Morden Wharf, East Greenwich, June 1854

Birth of the Industry 1720 - 1856

With the land lines also in place on the Irish side, the thoughts of the "projectors" on both sides of the Atlantic could now turn to the vast undertaking of bridging the ocean that separated them.

In July 1856, Cyrus Field was despatched for the third time to England and once again met John Watkins Brett and Charles Tilston Bright with whom he had been corresponding since his previous visit. They agreed that a British company would be required in order to secure the maximum interest from British investors. So all three put their names to an historic agreement dated 29th September 1856:

> ... to exert ourselves with the view, and for the purpose of, forming a company for establishing and working of electric telegraphic communication between Newfoundland and Ireland, such Company to be called the "Atlantic Telegraph Company"...

On 3rd October, the Magnetic laid on yet another nocturnal demonstration, this time for Professor Samuel Morse who was electrician to the New York, Newfoundland and London Telegraph Company, and in the early hours of the following morning Morse wrote to Field:

> The experiments have most satisfactorily resolved all doubts of the practicability as well as practicality of operating the telegraph from Newfoundland to Ireland.

The Atlantic Telegraph Company was registered on 20th October 1856. With a solid agreement in place and the serious technical questions at least partially answered, the great enterprise could at last begin.

Left
The signatures of Cyrus Field,
John Watkins Brett and
Charles Tilston Bright agreeing to
form the Atlantic Telegraph Company

Above: HMS Agamemnon at Greenwich loading the 1857 Atlantic Cable

"Throwing a Girdle Round the Earth"

Section 2 The Telegraph Era 1856 - 1956

"The confidence that Victorian Britain felt in its technology, in its commerce and its political system could hardly have been better expressed than through the development of submarine cables. The technical skills and the capital were obtained from Britain, and Britain was the centre from which the lines reached out to embrace the globe. Distances from London that had formerly been measured in weeks or months were encompassed in minutes, with profound consequences for politics and commerce.

A special technology was developed with the cables, including unique methods of signalling and detecting that were necessary because of the long distances covered and the electrical characteristics of the lines. This technology, or at least a large part of it, became obsolete in the 1950s and 1960s when amplifiers were developed which could be submerged with the cables and when communications satellites began to appear. Old cables, old instruments, old technology were abandoned."

"Submarine Telegraphy,
The Grand Victorian Technology"
by Bernard S Finn, 1973

The Great Enterprise Begins

The first great adventure in the story of long distance submarine cable communications was the laying of the cable across the Atlantic. This has since been described as the Victorian equivalent of the Apollo moon landing project, due to its great cost and technological sophistication. This parallel gives us a good sense of the amazing nature of this project and its world-changing effect.

As with most major capital undertakings, the key incentives for laying an Atlantic telegraph cable were economic. The growing export industry of the 1850s in Britain made it ripe for improved international communications. This boom helped to encourage industrialists and entrepreneurs with access to capital and technology, turn their visions into reality. The London banker needed to know the state of the stock exchanges in Paris and New York. The Liverpool shipper gained considerably if he could communicate with his ship's captain in Singapore about the commodities available and their prices. The atmosphere was conducive to change and development of the most radical kind. The Atlantic cable was one such radical idea.

Following the formation of the Atlantic Telegraph Company, activity was intense. Firstly, the £350,000 required to fund the project had to be raised. No prospectus was published, no advertisements were made and, at the outset, there was no Board of Directors or officers of the company. The fundraising began in Liverpool on 12th November 1856 following the issue of a short notice by Mr Edward Bright. This notice brought the scheme to the attention of the foremost traders and businessmen of the city. Cyrus Field, John Brett and Charles Tilston Bright spoke and inspired the audience to hand over their money! The first subscribers were Charles Tilston Bright, the Mayor of Liverpool and Mr Charles Pickering of Messrs Schroder. Many others soon followed them. Similar meetings were held in Manchester, Glasgow and London. Within just a few days the £350,0000 had been raised through the issue of 350 shares of £1,000 each. The shareholders included George Peabody, better known today as the philanthropist who provided (and who through his trust still provides) good public housing in London. Two other names stand out because of the influence they were to have on the future development in international communications. One was the Manchester-based cotton merchant and entrepreneur John Pender, who was already a Director of the Magnetic Company. The other was Professor William Thomson of Glasgow University. His work on the operating equipment for submarine cables was key to the development of the industry.

A Glorious Failure

The construction of the Atlantic cable began in February 1857. The job of making the core was given to the Gutta Percha Company of London. The outer sheathing of the cable was divided equally between Glass, Elliot & Company of Greenwich and R S Newall & Company of Birkenhead. This was a practical solution as the cable was to be laid in two main sections by two ships. The process involved 119.5 tons of copper drawn out into 17,500 miles of wire. Seven strands of this wire were combined to make the core 2,500 miles long and 300 tons of Gutta Percha were used for the insulation. To strengthen the cable, 315,000 miles of outer sheathing. The total mileage of wire made for the project was more than enough to have gone around the earth 8 times, and all in 4 months!

The ships used to lay the cable of 1857 were the finest available. The British ship was HMS *Agamemnon*. At 3,500 tons, she had a huge hold of 45 feet square and 20 feet deep suitable for holding nearly half of the cable, with the rest divided into two smaller coils. The other half of the cable, made at Birkenhead, was coiled on board the new 5,540 ton American steam frigate USS *Niagara*, captained by W I Hudson. Both ships were fitted with specially devised machinery for paying out the cable.

Above
USS Niagara, with HMS Agamemnon at
Keyham, re-loading for the 1858
Atlantic expedition

The Telegraph Era 1856 - 1956

The plan was to start the lay from mid-ocean but this was changed to a start from Ireland at the insistence of Dr Edward Whitehouse the electrician. So, on 7th August 1857 the *Niagara* began laying the cable from Valentia Island, off the far-west of County Kerry in Ireland. Trouble started early when the cable broke after only three-quarters of an hour so they started again, only to break the cable again after 334 nautical miles had been laid. The ships returned to port and were unloaded, and 900 hundred miles of new cable were ordered for another attempt the following year.

On 19th June 1858, the two ships set out again with 3,000 miles of cable between them. This time, despite the *Agamemnon* nearly sinking in a massive storm on the way, they met in mid-Atlantic and after splicing the cable together, set off on 26th June in opposite directions to lay the cable. After three breaks, the ships once again returned to port. Fortunately, it was discovered that there was still enough cable left to make one last attempt so they returned to mid-ocean and began laying again on 29th July. Finally, on August 5th both ships arrived at their respective destinations, Valentia and Trinity Bay Newfoundland. The cable was laid! On 13th August, Queen Victoria and President Buchanan exchanged messages over the cable to mark its formal inauguration. There were widespread celebrations in America and somewhat more muted ones in Britain; Charles Tilston Bright was knighted for his role as project engineer. However, the cable never opened to commercial traffic, because by September, it had become difficult to get any clear reading from the receiving apparatus and after 20th October it was not possible to send a signal of any kind. The reasons for the failure of the cable, as far as could be ascertained, were damage it had sustained in transit and the application of huge 3,000 volt pulses required to operate the standard receiving equipment used.

Cyrus Field and his colleagues were not put off by the failure of the 1858 cable. They had demonstrated that it was possible to lay a cable under the Atlantic and to send messages along it, they just needed to improve the technology involved. However, the American Civil War and another major cable-laying disappointment in 1859 with the failure of the Suez to India cable at a loss of £800,000, made many investors rather wary of sinking any more of their money in this new and so far, unproven technology.

The Industry Advances

The particular electrical characteristics of a long undersea telegraph cable meant that it was vital that someone invented an effective device for receiving the faint signals which made their way to the other end. The two main problems were the feebleness of the signal received and the fact that the cable stored electricity (capacitance) as well as transmitting it, which caused a blurring of the signals. This meant that signals had to be sent very slowly to stop pulses from merging into one another. Later, measures were taken to stop this, enabling great increases in the speed of signaling.

The man whose ingenuity solved these problems in the early days was William Thomson (1824-1907), professor of Natural Philosophy at the University of Glasgow and an adviser on the Atlantic cable project. He designed many instruments which were key to the development of international communications. One of these inventions was the Mirror Galvanometer, the first

31

Right
*Brunel's huge ship, SS Great Eastern
laying the Atlantic cable*

receiving device with sufficient sensitivity to detect the tiny attenuated electrical signals coming over early long distance cables. This was essentially a very light mirror and magnet suspended at the centre of a coil. The received signal from the cable was directed through the coil, creating a very small magnetic field, which turned the magnet and mirror. A beam of light shone onto the mirror and this in turn deflected onto a scale so that when the signal came through, the mirror moved slightly and showed a corresponding movement of the light beam on the scale.

The cable industry had come a long way since the early 1850s. R S Newall & Co had a quiet year in 1856, making only three cables. However, the 1857 order for half of the Atlantic cable improved things significantly. Newall's was also commissioned to manufacture and lay cables in the Mediterranean between Sardinia, Malta and Corfu in 1857. In 1858, they made and laid cables in Greece and were also awarded a contract for cables which they part funded, laid by the Red Sea and India Telegraph Company across the Suez isthmus and down the Red Sea. All the concessions granted to this company eventually were transferred to the British Indian Submarine Telegraph Company. The main feature of Newall's cables was that they were lightly armoured and tended to fail after short periods. Between 1860 and 1866 they did not make any cables at all but were later commissioned by the Danish-Norwegian-English Telegraph Company (which would become part of Great Northern) to make and lay a rubber core cable. They also supplied the cable linking Cornwall with the Isles of Scilly in 1869 and a duplicate Scotland-Ireland cable (Portpatrick to Donaghadee) in 1870. After this, the company ceased to make submarine cables. With R S Newall's departure from the scene

almost the entire world's submarine cable manufacturing capability was based on the River Thames in London. On the north side of the river there was Hooper's Telegraph Works Ltd at Millwall, The India Rubber, Gutta Percha and Telegraph Works Co at Silvertown and W T Henley Telegraph Works Co at North Woolwich. On the south side was Siemens Brothers at Charlton and at Greenwich the company that would dwarf them all.

On 7th April 1864, a new cable-making and laying company was founded, the Telegraph Construction & Maintenance Company (Telcon). This was a merger of the Gutta Percha Company and Glass, Elliot & Company. The first chairman of the new company was John Pender.

An Ocean Tamed

One of the first actions of Telcon was to approach the Atlantic Telegraph Company with an offer to make and lay a new cable across the Atlantic. John Pender and Daniel Gooch raised most of the £500,000 necessary capital and chartered Brunel's huge ship, the *Great Eastern*, to lay the cable. The shore end cables were made by W T Henley in North Woolwich. The *Great Eastern* left Valentia on 23rd July 1865 and paid out 1,186 nautical miles of cable until, in an attempt to correct a fault in the cable, it broke and was lost. This near success led to another attempt. An additional £600,000, was raised by Gooch and Pender by establishing a new limited liability company, the Anglo-American Telegraph Company, and on Friday 13th July 1866 the *Great Eastern* set out again. This time the cable was successfully laid all the way to Heart's Content, Newfoundland, arriving on 26th July. The *Great Eastern* then returned to the location of the lost 1865 cable and, after many attempts,

Right
Cable being landed at Porthcurno, Cornwall

managed to retrieve it. A new end was spliced on and by 8th September there were two new Atlantic cables. This brought the personal crusade of Cyrus Field to a successful conclusion, and his untiring efforts were properly recognised in the United States, by a unanimous "Vote of Thanks" in Congress. In Britain, the success was recognised by Queen Victoria, Daniel Gooch received a baronetcy (Gooch of Clewer), William Thomson was knighted, as were Samuel Canning and Richard Atwood Glass, Chief Engineer and Managing Director of Telcon respectively. Surprisingly John Pender was not honoured even though he had risked his entire fortune to under-write the successful project.

Improvements in System Design

Cables continued to be made in more or less the same way for the next 100 years. Broadly speaking, they were composed of many layers of different conducting, insulating and strengthening materials. In the centre there was a core of copper wire covered in a layer of Gutta Percha. This was followed by a layer of jute with a covering of steel armouring wires. Further layers of jute were then applied and finally the cable was dipped in a protective compound. This type of cable was used for many years. The thickness of the steel armouring wires differed, depending on whether the cable was for use in shallow or deep water. In shallow water the armouring was thick and heavy to protect the cable from damage on rocks or by anchors of nearby boats. As the cable went into deeper waters, there was less need for heavy armouring and the wires were not so thick.

There were however some important developments in cable design and capacity. After 1879 brass tape was often added around the insulated core to protect it from boring insects which liked to eat the Gutta Percha. In terms of cable capacity and the speed of signaling, in the 1920s an iron-nickel alloy called Permalloy was produced by Western Electric and a copper-iron-nickel alloy called Mu-metal, was produced by Telcon. This was incorporated into the cable design to "load" the cable with inductance so that the signals would not become so distorted by the capacitance of the cable. The first "loaded" cable, using a tapered loading Mu-metal design, was laid in the Atlantic in 1924 and achieved five channels, each at 50 words per minute in both directions simultaneously an increase of over five times the unloaded design capability. Many other loaded telegraph cables were to follow across the Atlantic and elsewhere in the world. The last to cross the Atlantic was laid by Telcon for the Western Union Telegraph Company in 1928, between the Azores and Newfoundland.

By the 1930s there were just two firms making submarine cables in Britain, Telcon and Siemens Brothers. With the depression and competition from radio, business became more and more difficult to obtain and in 1935 there was a merger of Telcon with the submarine communications cables section of Siemens Brothers to form Submarine Cables Limited.

Transmitting and receiving equipment also evolved slowly over the years. Special keys were designed for Curb transmission, described by William Thomson and Fleeming Jenkin in an 1861 patent. Curbing employs a reverse pulse transmitted immediately after the main pulse to help sharpen the received signal. In the 1880s, automatic machines fed by punched tape

were introduced. At first, the punched tape was produced manually but after 1900, it was produced by keyboard perforators, which were operated in a manner identical to an ordinary typewriter. Problems at the receiving end were more complex. William Thomson designed a siphon recorder, which was less sensitive than the mirror galvanometer but had the advantage of creating a permanent record of the received signal on a narrow strip of paper or "slip". The siphon recorder was used from around 1870, with ten being specially made for the new England to Bombay route, which was completed in that year by John Pender's companies.

All this technology still required expensive human input to transmit the tape, read the message at the other end and re-transmit it on to the next cable on the route if necessary. With a view to economy of operation, a great deal of effort was put into finding a technology which could cut out the middle man and automatically receive and retransmit telegraph messages. To do this the signal first had to be magnified at the receiving station. One method of magnifying, was invented by E S Heurtley, in 1908. It consisted of a moving coil similar to that of a siphon recorder, which was attached to two wires. The wires were heated by the electric current flowing through them and cooled by two small blowers. When the coil turned, one wire moved closer to its blower, the other moved away. Since resistance depends on temperature, the resistance of the first wire decreased and that of the other increased. This upset the balance of the electrical circuit of which the two wires were a part and produced an amplified signal, large enough to operate an electro-mechanical relay and generate a new pulse.

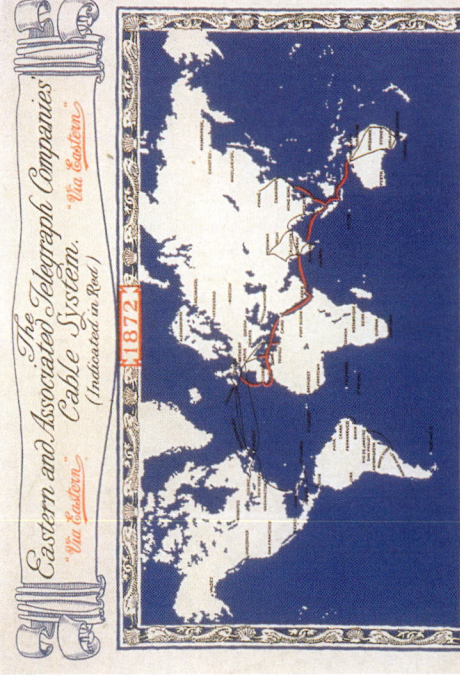

Above
A map of the Eastern and Associated Telegraph Companies showing the extent of their operations in 1872

Another important development in cable capacity was the introduction of duplex signalling. This simply meant sending signals down one telegraph wire in opposite directions simultaneously. This began as early as the 1850s on overland telegraph wires but it was a much more difficult task to achieve on submarine cables because of their capacitance. However, duplex working was achieved successfully in the 1870s using, initially, a method developed by J B Stearns, and then more effectively in 1875, by using H A Taylor and Alexander Muirhead's method which employed an "artificial line" to duplicate the electrical characteristics of the cable. At the transmitting end of the cable the signal would be sent down the real cable and the artificial line simultaneously. The effect of this was that the local receiver would only see the incoming messages.

An Eastern Empire

With the success of his investment in the Atlantic cable, John Pender saw great opportunities in laying cables along other routes, where communications were vital. In particular, he foresaw the importance of communications with India for government and trade purposes. In 1868, Sir Daniel Gooch replaced John Pender as Chairman of Telcon as Pender devoted more time to his grand plan. The first order for this new network received by Telcon was for the Malta to Alexandria cable of 1868 and the success of the company thereafter was based largely on Pender placing orders with them for the cables which linked the British Empire.

In 1863, the British Government had sponsored the laying of a cable in the Persian Gulf from Karachi to Fayal. This was linked to Europe via land lines through Persia and Turkey. Charles Tilston Bright was the engineer on this project and the cable was successfully completed in 1864. The main problem with this line to the East was the landlines, which proved unreliable and passed through the territories of various tribes who had little respect for the telegraph wire. Again, these flaws in the quality of service encouraged people such as John Pender to further pursue the idea of an all submarine cable network, which could not be interrupted by local problems. In 1868, John Pender set about forming a number of telegraph companies to lay an all-submarine route to India, which by-passed these problems and by 1870 the link was complete. The UK landing point for this cable chain was Porthcurno in the far-west of Cornwall. Also in 1870, UK inland telegraphs were finally nationalised, releasing private capital and boosting investment in international telegraphs.

By 1872, the Far East and Australia were also linked up to the network and on 1st June of the same year, John Pender merged his group of companies to form one large company under the name of "The Eastern Telegraph Company Limited". In 1872, the Eastern Telegraph Company owned 8,860 miles of submarine telegraph cables, owned or rented 1,200 miles of landline, had 24 stations and two cable repair ships. Its revenue for the year was £376,900. The cost of sending a telegram from England to India in "Via Eastern" was £4 per message, prohibitively expensive for anyone other than government and large businesses. However, for those who could afford it, the telegraph revolutionised the speed of business and greatly improved the effectiveness of British Government around the Empire. The service from London to Bombay and beyond was very efficient; the sender of a message from London to India in 1872 could expect a reply within 24 hours.

Under John Pender's guidance, the Eastern Telegraph Company grew and grew, duplicating and even triplicating cables on the busiest routes. It was not long before the West Indies, South America and Africa were the scene of feverish cable-laying activity by many new companies and one-by-one, these small companies were taken over by the Eastern Telegraph Company and became part of the company's international network. By 1887, the Eastern owned 22,400 miles of cable, had 64 stations, a capital of £5,900,000 and gross annual revenue of £650,971. In recognition of the widening geographical coverage to the West as well as the East, the company adopted the name Eastern and Associated Telegraph Companies. To its employees it continued to be known as "Father Eastern". John Pender's contributions to the Empire were finally recognised in January 1888 when he was knighted KCMG. This honour was raised to GCMG on 29th June 1893.

By the turn of the 20th century, John Pender's company had created the world's largest international telecommunications system, linking Britain with the Empire and most of the world by an almost totally submarine network. The effect on the world of this and the other communications networks was immense. Whether the cables inspired greater nationalism or whether they were a vehicle for peace is debatable but there is no doubt that the increased speed of communication radically affected the speed of political negotiation as well as of war. The effect of the telegraph cables on trade was clear, as were the effects on the communication of press information. Public awareness and reaction to world events was made possible via this vast communications network.

Cableships and Surveys

This global network was laid by an ever-growing fleet of specialist ships. The *Hooper*, launched in 1873, was the first purpose-built cable-laying ship and at the time the second largest ship afloat. Prior to this as we have seen, the largest ship the *Great Eastern*, and others were converted to enable them to undertake cable-laying. In 1881, the *Hooper* was acquired by the India Rubber, Gutta Percha and Telegraph Works Company, and renamed *Silvertown*. The unique identifying features of a cableship were, and still are, its tanks for storage of the cable and the cable handling gear inside and on the deck and bow or stern sheaves. To the casual observer, the most obvious external indication that a ship is involved in cable work are the bow or stern sheaves used to pass the cable out of, or into, the vessel.

At a very early stage, fishing vessels were catching their equipment on submarine cables and were a major cause of faults. It was often difficult to locate these faults and further delays were caused by the non-availability of suitable vessels to carry out the repairs. It was therefore essential for system owners to have access to specialist cableships for repairs. The British Government, therefore decided to have a new dedicated cableship built and the work was authorised by Parliament in December 1881. The vessel was built by Dunlop & Company of Port Glasgow and named *Monarch (II)* when launched on 21st August 1883. By the late 1880s, agreement was reached between the British Government and various Continental governments to take a joint interest in the submarine cables connecting their countries, and henceforth these cables were maintained at joint expense. This collaboration was the

The Telegraph Era 1856 - 1956

Above
HC Oersted, the first purpose built cable repair ship

beginning of the now well established, maintenance zones which cover all of the world's oceans. Within these zones cable owners share the burden of keeping specialist ships on 24-hour stand-by in order to respond quickly to cable faults. The requirements of cable-laying and cable repair in many areas are quite different and therefore two types of cableship emerged, laying ships owned and operated by the cable manufacturers and repair ships owned and operated by the cable owner.

In addition to creating communications links, cable-laying had a significant effect on the understanding of the world's oceans. The ocean beds of the world are as varied as dry land but until cable-laying began there was very little information about them. In the 19th century, the limited knowledge of the ocean floor that did exist was often the result of soundings taken in the course of laying telegraph cables. In fact, the names of cableships have been given to many features of the ocean beds of the world. The Atlantic, for example, has a huge ridge running down its middle that includes the "Faraday Hills", discovered accidentally by the cableship *Faraday*. These reach 6,000 feet high, higher than the tallest mountain of the British Isles. The cables laid on the bottom of these oceans have always been vulnerable to damage. In the case of telegraph cables, there were worms that burrowed into the cable and ate the Gutta Percha insulation. Rough sea-beds, underwater volcanoes and strong currents can also cause damage if the route is not carefully planned to avoid them.

The Danes Go East

John Pender's dominance of telegraph communications to the Far East, was not seriously threatened by any company; apart from Denmark's Great Northern Telegraph Company. On 1st June 1869, Great Northern, was formed by a merger of, the Danish-Norwegian-English Telegraph Company, the Danish-Russian Telegraph Company and the Norwegian-English Telegraph Company. During 1868 and 1869, using cables made by R S Newall and W T Henley, these companies established links between England, Scandinavia and Russia. The driving force behind Great Northern was Carl Frederik Tietgen, who also formed the Great Northern China and Japan Extension Telegraph Company, on 9th January 1870. This company awarded a contract to provide cables between Vladivostok, Shanghai, Hong Kong and Nagasaki to Hooper's Telegraph Works, who used Siemens Brothers to armour the cables. By 1st January 1872, a link had been established from England to China, Japan and Hong Kong via the Great Northern's European network and the Russian landline, linking with this Far East cable network. Great Northern's advantage over Eastern was the fact that their route was significantly shorter and therefore quicker than the all marine route. This commercial advantage was entirely due to the exclusive cable landing rights that Tietgen had negotiated with the Tzar. To maintain their eastern cables, in 1872, Great Northern commissioned the first purpose-built cable repair ship, *H C Oersted*, named after the discoverer of electromagnetism.

One of the main problems faced by the Great Northern and caused by its success in the Far East, was how to telegraph Chinese characters. This they solved by the creation of a special code in which the most frequently occurring characters were each assigned to a four figure number group. Their code remained in use for over 100 years.

In 1883, shortly after the Eastern Extension Company had laid their cable from Hong Kong to Foochow and Shanghai, the two companies entered into a "joint purse" agreement. The extent of co-operation was surprising, they shared offices and even sales desks. In 1900 and 1901 a further six cables were laid in the region by the two companies but in 1917, Great Northern's network was seriously threatened by the Russian Revolution. Amazingly, they were able to re-negotiate and in July 1921 their concession was renewed, when Vladimir Ilich Lenin signed a new agreement in the Kremlin. Great Northern continued to prosper until the outbreak of the Second World War, when all their vital marine cables were cut. During the war many of their key concessions expired and while they did re-open the Vladivostok - Nagasaki route they never got back into China and were unable to re-establish their leading position.

The Last Great Challenge

By the final decade of the 19th century, the Eastern and Associated Telegraph Companies had become hugely successful and had a monopoly on international communications in many parts of the world. One area in which they could be challenged was the absence of a cable across the Pacific, this was originally mooted in 1877, but became the subject of great discussion from 1880 onwards. The subject was raised by Australia, which objected to Pender's stranglehold on communications to and from their continent. The first significant move to undo this was taken in 1893 when a concession was awarded to a French company to operate a cable from Australia to the French island colony of New Caledonia. The contract for the cable was awarded to the Société Industrielle des Téléphones (SIT). Submarine cable manufacture had been established in France in 1891 with the building of a factory at Calais by the Société Générale des Téléphones (SGT) to armour core made at their Bezons factory. This company could not supply a complete solution and remained dependant on Britain for many aspects of a project. So, in 1893, SIT was set up absorbing SGT and purchasing their own cableship the *Westmeath* which they renamed the *Francois Arago*. This was the start of the submarine cable industry in France and the first real challenge to the virtual monopoly of the British.

The New Caledonia project was subsidised by the New South Wales Government and the French Government and it worried John Pender greatly. He saw it as a precursor to a French contract for a Pacific cable which would operate in competition with the Eastern Extension Company and reduce his revenue.

The Telegraph Era 1856 - 1956

Negative press was generated by Eastern in an attempt to put people off the idea of a Pacific cable at all. France, of course, had a particular interest in Canada which had formerly been one of its colonies and where there were still many French people living. By 1892 the British Admiralty had surveyed the proposed route and in 1893 the Canadian Minister of Trade travelled to Australia to present a detailed plan for the Pacific cable prepared by Mr Sandford Fleming. He argued that if economy, low rates and high efficiency were required then the project should be carried out under government control, not by the Eastern and Associated Telegraph Companies or any other commercial company. He proposed that Australia, New Zealand, Fiji and Canada should be the joint owners of the cable and that the cable should be operated as a public service. These proposals were met with apparent indifference by most, as people were still loath to publicly oppose the Eastern Extension Company and face the possibility of increased telegraph rates. So, negotiations went on at a snail's pace until Joseph Chamberlain took up Canada's suggestion that an Imperial Pacific Cable Committee should be formed to examine the project. This met in London on 5th June 1896 and worked until 12th November interviewing numerous "witnesses". Among these were the Eastern and Associated Telegraph Companies of course, this time led by a new Chairman, the Marquis of Tweeddale. They were still objecting to the Pacific cable project and still maintaining that their lines between Darwin, Penang and Madras were quite adequate to carry all the telegraph traffic to and from that region. The report of the committee was not published until April 1899 but after much further prevarication, the project was finally put into action. Telcon received the contract for making and laying the cable on 31st December 1900. This was estimated to be 8,000 miles long in five sections. The routes were: from Doubtless Bay on the north-western tip of New Zealand to Norfolk Island, 500 miles to the north-west; from Southport, just south of Brisbane, Queensland, also to Norfolk Island; from Norfolk Island to Suva, Fiji; from Suva to Fanning Island; and from Fanning Island to Bamfield on Vancouver Island, Canada. This last section, at 3,458 nautical miles, was the longest telegraph cable ever laid. It was laid by the *Colonia*, a ship built specially by Telcon for the purpose. She had four cable tanks with a total loading capacity of 143,962 cubic feet. The Pacific Cable Board also had the *Iris* built in 1902 to maintain the cable.

When the bill to provide the funds for this project came for its second reading in parliament on 12th August 1901, one John Henniker Heaton, an Australian newspaper owner and MP for Canterbury, attacked the conduct of the Eastern Telegraph Company:

Above
The cable laying ship Colonia

The Telegraph Era 1856 - 1956

"I know no monopoly in the world that is doing more injury to trade than the concentrated companies represented by the Eastern Telegraph Company and its six or seven satellites.... I once described John Pender and Co. as an octopus which, with its tentacles in every direction, is sucking the lifeblood out of the Empire.... I regard the scheme now before the House as a great step forward towards the breaking up of one of the greatest monopolies the world has ever seen and towards the consolidation of the Empire."

Harsh words, perhaps, but nonetheless words which clearly demonstrate the perceived power of the Eastern Telegraph Company and the influence of its telegraph network on the workings of world business.

The Pacific Cable Act was finally passed on 17th August 1901 and laying began in 1902. The first message was sent on the 31st October and the opening of the cable caused as strong a reaction as that generated by the Atlantic cable nearly 40 years earlier.

The Pacific cable was laid at around the time when "multiplexing" appeared on busier telegraph routes. By the 1920s, mechanical signalling had achieved speeds of over 200 words per minute compared to the 12 words per minute of the 1866 Atlantic cable. In 1926 the Bamfield-Fanning-Suva sections of the Pacific cable were replaced with loaded cables, enabling a significant increase in traffic. A much larger cableship (9,073 tons) named the *Dominia*, was built by Telcon for this installation. The Pacific Cable Board remained as a separate entity operating these cables until 1928.

An Air-borne Attack

The Pacific cable had been completed at a time when another new international communications technology, radio, was in its infancy. There were some things that the powerful Eastern Telegraph Company and its rivals could not yet offer its customers, namely inter-continental voice communication. It was not yet technically possible to send a telephone message over a long undersea cable, due to the fact that the frequency range required for even a single voice was far beyond the transmission capabilities of technology which conveyed telegraph signals at speeds of only 50 pulses per second or less. There was also need for ship-to-shore communications which cables could not provide.

The technology which was first to make long distance voice communication possible and which provided ship-to-shore contact, was radio. In 1895, the Italian physicist Guglielmo Marconi began to make practical steps forward in his experiments and in 1896 he patented radio communication. He conducted most of his experiments in, or off the coast of England and in 1901 he succeeded in signalling a Morse Code message from Poldhu in Cornwall to Newfoundland, 2,000 miles away across the Atlantic. Eastern Telegraph staff at Porthcurno could see the aerial wires which Marconi had erected and wondered what was going on but certainly saw no real threat to the cable system at that time. However, in 1902, they erected a small hut on the cliffs and set up some rudimentary radio equipment with the intention of spying on Marconi's activity at Poldhu.

*Left
Receiving room, Marconi
radio station, Glace Bay
Marconi seated to the left).*

wave systems. The government was impressed and by 1927, Post Office Beam Wireless telegraph services had been established with Canada, Australia, South Africa and India. On the night of 14th January 1923, another major breakthrough occurred with the first transatlantic radio telephone call from the RCA transmitter at Rocky Point, New York to STC's New Southgate factory, albeit a one way event, the die was cast. By 1926, STC had built a transmitter for the Post Office at Rugby and on 7th January 1927 a transatlantic telephone service was open to the public. A three-minute call costing the princely sum of £15. For the first time, the cable companies began to feel the effects of radio competition with a significant loss of revenue.

The Rivals Come Together

This threat to the cable systems resulted in a call for government action to prevent the huge cable infrastructure from foundering and in 1928 the Imperial Wireless and Cable Conference was convened to "examine the situation which has arisen as a result of the competition of the beam wireless with the cable services . . . . etc." The outcome of the conference was a recommendation that British national and imperial interests would be best served by a unified system of cable and wireless communications. During September 1929, a new operating company was formed with the name "Imperial and International Communications Ltd" to take over the assets of the Eastern and Associated Telegraph Companies, the Pacific Cable Board and Marconi's beam radio network. This company was renamed Cable & Wireless in 1934.

The combined cable and radio telegraph network went on to play a vital strategic role during the Second World War. The

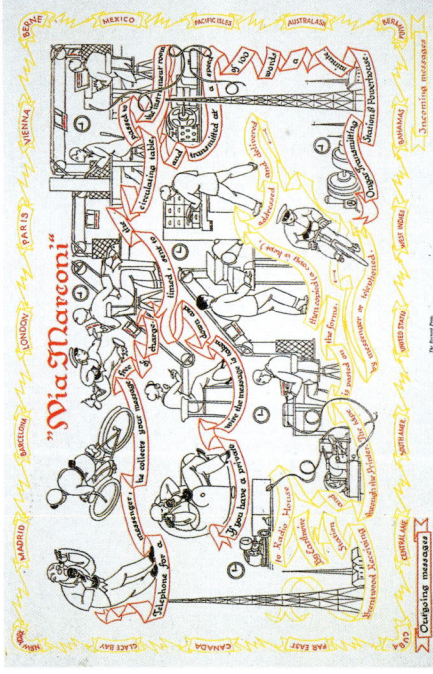

Above
An advertisement for Marconi's Wireless Telegraph service

The threat of radio increased with Marconi's technical advances. In 1907, Marconi opened a radio-telegraph service between Clifden in Ireland and Glace Bay in Canada. This was followed by other services, including voice communication by radio-telephone to Europe, the USA and Australia. He used low frequency long-waves, which required expensive aerials and high-power transmitters.

Much of Marconi's further development work was carried out on board his private yacht *Elettra* which he bought in 1919. He used the ship as a mobile laboratory and went on to develop a short wave "beam" radio system and demonstrate it to the British Government. He planned to build an Imperial Wireless telegraph chain which would transmit at three times the speed, use 2% of the power and cost about 5% of the existing long-

The Telegraph Era 1856 - 1956

Above
"Zodiac" Eastern and Associated Telegraph Company's in-house magazine

speed at which it was fought was far greater than any previous war: campaigns previously two years long could be completed in around six weeks thanks to the speed of communications. Radio was necessary for the tactical handling of troops but cables were essential for strategy as they provided secure private communications whereas radio could be intercepted. It was fortunate for Britain that, at the outset of war, Cable & Wireless owned 155,000 of the world's 350,000 miles of cable as well as 91 wireless circuits covering over 200,000 miles of the earth's surface.

During the Second World War there was an overall increase in telegraph traffic of about 400%, much of it military. Despite the fact that vital communications were being carried by what remained an independent company, the British government were happy to allow Cable & Wireless to remain independent during the war. In view of the strategic importance of the network, this has been described as "the greatest compliment ever paid to private enterprise". Although 700 staff were called away on military service, the technical staff were compulsorily frozen in their civilian occupations.

After the war, a decision was made by the Labour government to bring Cable & Wireless into public ownership, although it was not until 1st January 1947 that nationalisation actually took effect. This brought to an end four generations of stewardship of the company by the Pender family. In many other ways nationalisation did not dramatically affect the operations of the company, most of which were outside Britain. The priority for everyone was to restore the cable and radio routes damaged during the war.

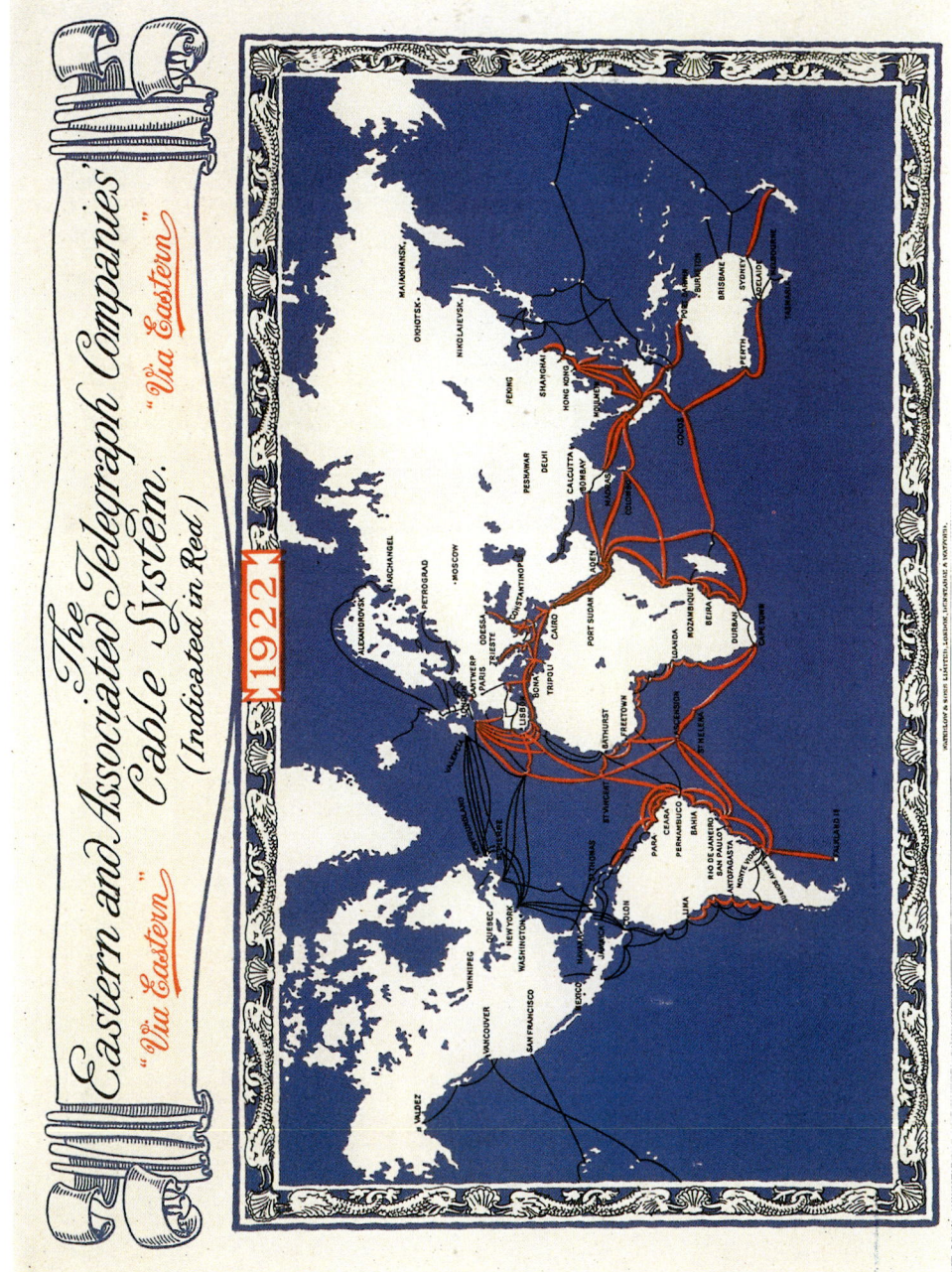

Above
1922 map of the Eastern and Associated Companies showing the extent of their operations

The Telegraph Era 1856 - 1956

The World Begins to Change

As the centenary of the submarine cable industry dawned the British dominance of the manufacturing industry remained intact. The dynasty founded by Pender and Gooch was still the industry leader. By 1950, Telcon had manufactured 385,000 nautical miles of submarine cable, some 82% of the total market. The fast growing French industry had suffered a severe set back in 1940 when the Calais factory was destroyed in an air raid and they were just beginning to recover. There was still no serious manufacturing capability in the USA: the Simplex factory in Newington, New Hampshire would not open until November 1953. The only other manufacturers in Germany and Japan had been devastated by the Second World War. The German industry, which dates back to 1901 and Norddeutsche, was worst hit and it would be many years before Germany could produce a viable submarine cable again. The Japanese industry had been triggered by the arrival of the Great Northern telegraph cable of 1871. However, the first cable manufactured in Japan was not laid until 1874, from then on, its industry was almost totally, inwardly focused, connecting the 3,900 islands of the Japanese Empire. After the Second World War a new factory was set up in Yokohama by the Ocean Cable Company to make co-axial cable. The competition was beginning to gather!

By 1955 there were 6,483,040 telephone subscribers in Great Britain and 52,806,476 in the USA, yet still they were unable to communicate with each other via an undersea telephone cable. The major technological advance, which changed this was the development of a new type of cable system, which enabled voice communication over long distances. This new "co-axial" cable design made of a new material developed by ICI, had signal boosters placed at regular intervals along it. These boosters were a joint product of Bell Telephone Laboratories in the USA and the British Post Office. As early as 1865 the first patent was taken out for inserting a device in the middle of a long telegraph cable to boost the signal current. However, it was not until 1950 that the Western Union Telegraph Company developed a thermionic valve amplifier to *repeat* telegraph signals. They inserted this in a cable from America to the UK laid in 1881. The device was christened a *"repeater"* and the name has remained with these submerged boosters to the present day, no matter what the technology.

With the advent of reliable submerged repeaters, a new era of communications was about to begin, one which saw the end of the Telegraph Era and the demise of the old cable networks that had helped to change the world so much during the previous 100 years. For the first time, voices would be heard under the great oceans.

An STC Repeater being laid

"And Nation shall Speak unto Nation"

Section 3 The Telephone Era 1956 - 1986

It's Good to Talk

Almost as soon as the fledgling submarine telegraph cable industry had established itself, visionaries were looking ahead to the next advance in technology. In 1854, the Frenchman Charles Bourseul, wrote in L'Illustration: "Whatever happens, it is clear that in the near or distant future the spoken word will be transmitted by electricity". However, this amazing insight took some years to come to fruition.

In 1860, a German school teacher, Philip Reiss, perfected equipment which enabled recorded music to be transmitted by electricity, along wires over long distances, he called his invention the *telephone*. However, the real breakthrough in speech transmission was achieved by Alexander Graham Bell on 2nd June 1875. While carrying out experiments on a system called the "Harmonic Telegraph", with the intention of increasing the traffic carrying capability of a single wire, he discovered, by accident, how a more complex signal could be transmitted down the wire using electricity. After further experiments he was granted a patent entitled "Improvements in telegraphy" on 3rd March 1876.

The telephone was an immediate success and networks developed quickly in many countries. These terrestrial telephone networks developed until, naturally, the question of their interconnection by submarine cable was raised. By 1891, the British Post Office had laid a telephone cable across the English Channel but its use was limited to relatively short distances due to the distorting effect of cable capacitance. It was Oliver Heaviside who first quantified the problems that had been observed in long telephone cables, and between 1885 and 1887 he showed that the distorting effect of the cable capacitance could be counteracted by adding inductance evenly spread along the length of the cable.

Fantastic Plastic

It was not until 1910 that the full importance of added inductance was realised and implemented in a cross-channel cable manufactured by Siemens Brothers. It was a great success and resulted in the immediate extension of telephony to the furthermost towns in France. Work continued on blends of Gutta Percha to reduce capacitance, and on the introduction of high permeability metals to increase inductance, but these were only able to achieve small improvements in performance. The big breakthrough came some years later.

In 1933, the laboratories of ICI made a key discovery, without which transoceanic telephony would never have been possible. Their discovery was Polyethylene, and this was better as a dielectric in every respect, than its' predecessor materials. It had a lower dielectric constant and lower losses; it was tougher, more easily processed, non-hygroscopic and cheaper. Polyethylene became available for experimental cable manufacture in 1938 and cables of this form contributed strongly to the war effort - the first one was laid between Cuckmere and Dieppe in 1945. After the war, Polyethylene and cable manufacturing factories once more became available for civilian purposes and a new Anglo-Dutch cable was laid in 1947. The design was very advanced for its time, armour protected with a 1.7-inch diameter air-and-polyethylene spaced co-axial core, it was able to carry 84 voice channels using radio frequency carrier transmission.

The problem now arose: how to combine the multiple voice-channel capability of the Anglo-Dutch cable with the much longer lengths required in order to span the Atlantic? From early studies, it seemed likely that submerged amplifiers and air-cored cable insulation would be required to counter and reduce the cable losses. A method of amplification, initially with thermionic valves and later with semi-conductors was required, and finally some way of housing and deploying these amplifiers on the ocean floor would need to be devised. As previously explained, these housings became known as repeaters.

An American proposal from 1942 had envisaged a repeater consisting of a hollow flexible steel envelope containing valves and other components. It was thought that using such repeaters at a 50-mile spacing would enable the provision of 12 transatlantic cable voice circuits at a price comparable with the same number of radio circuits. The first repeater in commercial service was designed, manufactured and installed by the British Post Office in 1943. It was connected into a previously laid Paragutta insulated co-axial cable linking Holyhead with the Isle of Man by the *Iris (II)* on 24th June. It was with this experience, and considerably more faith in researchers and designers, that the Post Office, the American Telephone & Telegraph Company, the Canadian Overseas Telecommunications Corporation and the Eastern Telephone and Telegraph Company (a Canadian subsidiary of AT&T) met and made an agreement to build the first Atlantic Telephone Cable TAT-1.

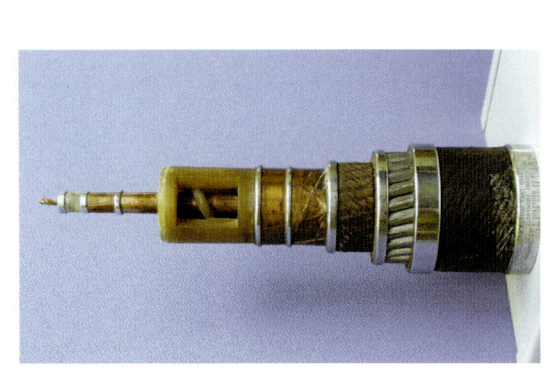

Right
Sample of 1947
Anglo-Dutch cable

Below
The world's first commercial repeater

The Telephone Era 1956 - 1986

Transatlantic Telephone One

Representatives of Bell Telephone Laboratories and the British Post Office immediately opened discussions on the technical form of such a cable. It emerged that American work dating back as far as 1919 had resulted in a thin, flexible repeater capable of passing through standard shipboard laying equipment. The design was capable of withstanding deep water and used a valve of very conservative design - unaltered since 1941. A system from Key West to Havana, installed in 1950, had given them considerable field experience of these repeaters. The system comprised a pair of uni-directional cables providing 24 two-way circuits and was considered a good prototype for a transatlantic system.

British developments to date had resulted in a large rigid repeater, which had more room inside for filters and could consequently operate bi-directionally on a single cable. The capacity of 60 voice circuits was higher than the American design, but the repeater could not pass through the cable-laying equipment and would mean stopping the forward movement of the cableship at each repeater. This presented a risk of loops forming in the cable whilst in deep water and it was this fear that led to the acceptance of the 73mm diameter American design. These could be "stretched" to 36 circuits of 4kHz each and laid in deep water without stopping. The Post Office repeater design was to be used to achieve single-cable operation across a land and (shallower) sea route from Clarenville, Newfoundland to Sydney Mines, Nova Scotia. A radio-relay link would carry the New York circuits to the Canada-USA border where they would join the existing carrier network at Portland.

Above
The British rigid repeater

London to Oban capacity would be provided on 12- and 24-circuit co-axial carrier cables. It is notable that, the requirement to link to other countries in Europe as well, makes this the first function of London as an international switching centre and required the system to be built as switched 4-wire circuits for the first time.

The transatlantic cable would run to the north of the 22 existing telegraph cables, without crossing them and would require 51 repeaters. The route from Clarenville to Sydney Mines required considerable investigation, to avoid the pitfalls of trawling activity and several cable crossings that would occur on the Newfoundland Banks. A route was finally selected which used numerous water sections in the sea, rivers and lakes, involved no crossing cables and was well inshore of the main trawling areas.

The cable design was a solid dielectric co-axial structure, with the strength provided in the external wires. These varied from 24 high tensile wires, 2.2mm diameter for the deep-water cable, to single and double armour using 8mm galvanised mild steel wires. The whole length of cable core was tested to 90kVdc; every joint was tested to 120kVdc and X-ray imaged; and for the first time, application of all the cable markings was controlled electronically. Cable length was measured in nautical miles, one cable nautical mile = 1.8553km. The cable was manufactured at three sites - 7,739km was made by Submarine Cables Ltd at Erith in Kent, 616km was made by the Simplex Wire and Cable Co in New Hampshire, USA and 111km for the land section across Newfoundland was made by Southern United Telephone Cables at Dagenham, Essex. The river Thames had to be dredged and new wharfs constructed at the Erith site to allow access for the largest cableship available.

Right
Section of TAT-1 shore end cable

Far right
HMTS Monarch (IV) loading cable at the Submarine Cables Limited factory at Erith, UK

The Telephone Era 1956 - 1986

It should be noted that the plan involved the insertion of many American repeaters into British cable and this precedent of "wet integration" would, in future, enable system purchasers to closely align the desired levels of national investment with equipment supply. This was also the forerunner of the Universal Joint.

The manufacturing of the repeaters required clean rooms on an unprecedented scale. The 102 American repeaters each contained about 60 components and were assembled at the Western Electric plant at Hillside, New Jersey. From there, they were sent to the Simplex plant for armouring. Each repeater assembly was about 25m long, including cable tails, and most of them were sent to England for jointing into the cable and loading into the cable tanks of the HMTS *Monarch (IV)*. The 16 British repeaters each contained about 300 components and were made in purpose-built "dairies" at Standard Telephones and Cables (STC) in North Woolwich. All components in the British repeater were made by or under supervision of the Post Office. In particular, the Post Office Research Branch Valve Group at Dollis Hill made the valves, the tail cables and cable glands were fitted to the repeater cases at Dollis Hill. After assembly, the repeaters were returned to North Woolwich for a three month monitored confidence test.

The terminal equipment was considered to be particularly "high-tech" at the time, although it looks very modest in comparison with today's technology. The equipment comprised about ten racks altogether at each end of each cable, including cable termination, duplicated power feed, transmission equipment and repeater supervisory system. Four sets of AT&T equipment were required for the two transatlantic cables, and two Post Office sets were required for the Clarenville to Sydney Mines section.

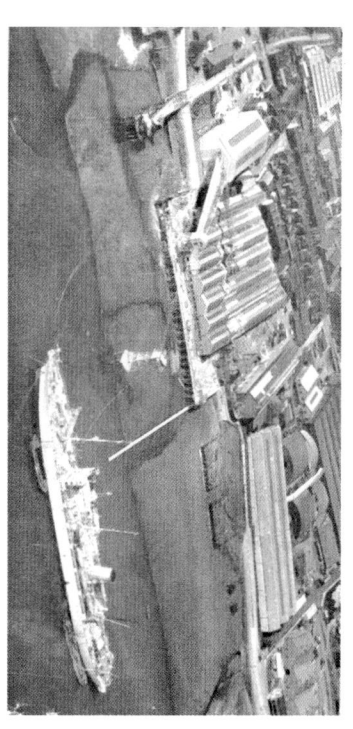

Captain W H Leech, OBE, DSC was the master of Her Majesty's Telegraph Ship *Monarch (IV)* for the installation and this cable lay will be a familiar story for today's cableship mariners. On her way to start the first lay, the *Monarch* collected cable and repeaters from two continents, took soundings to confirm the cable route and was delayed by storms and fog. Having laid the shore end from Clarenville, she returned to London to load the deep-water section of the first cable, but returned to the buoyed off end to discover thick fog, an iceberg and a cable end lost in 640m of water. The deep-water part of the first cable was completed and buoyed off at Rockall Bank, and more cable was loaded in London for the last leg to Oban. Unfortunately, the buoyed off end at Rockall was lost and storms with gales of 90 knots and 40ft seas delayed the recovery. However the job was completed on 26th September 1955 and the *Monarch* then went for her annual refit. On 1st February 1956, she started the program of load and lay for the Newfoundland to Nova Scotia segment, and in transit she ran into heavy seas and large areas of field ice off Penguin Island. Post-lay testing also revealed a faulty repeater, which was replaced. During June 1956, the second cable was laid from Oban to Rockall, and buoyed off. A second load, comprising the deep-water section, was laid without incident from Rockall as far as deep water near Clarenville by mid-July. The final shore end to Clarenville was laid, again without recorded incident, with the final splice "slipped" at 20:52 GMT on 14th August 1956 - which was 90 years and 18 days after the 1866 cable.

In the event of a cable break, traffic restoration was to be achieved over the existing radio circuits, and 12 additional operator positions were installed in London to cater for future growth. The traffic matrix for the system and its extensions was a good deal less complex than current systems, with circuits from the UK to the USA (22 circuits) and Canada (six circuits). Further circuits from the USA connected Germany (two circuits) with a single circuit to France, Belgium, Holland, Switzerland and Denmark. Norway and Sweden were connected to the USA indirectly via Denmark.

CANTAT

In May 1957, officials of the United Kingdom and Canadian Governments met in Montreal, where it was agreed in principle, to provide a direct telephone cable between the United Kingdom and Canada. The system, to be known as CANTAT, would be jointly owned by Cable & Wireless Ltd. and the Canadian

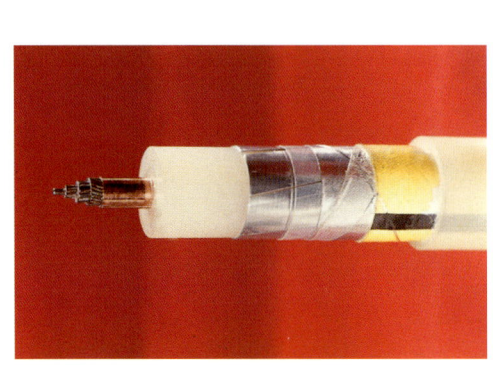

Right
A lightweight cable developed for CANTAT

The Telephone Era 1956 - 1986

Overseas Telecommunications Corporation, the project being formally announced in February 1958. A landing in Canada with no indigenous submarine cable industry opened the door once again for the whole contract to be delivered by a British company. The cable manufacture was in fact shared between Submarine Cables Ltd and STC, the latter having set up a cable factory in Southampton in 1956. The system contained two major innovations, bi-directional rigid repeaters deployed for the first time in a system of such length and lightweight cable, which was developed in conjunction with the Post Office Research department at Dollis Hill.

Almost all cables since the first successful submarine cable was laid across the English Channel in 1851 had an external covering of iron or steel armour wires. These wires served the dual purpose of providing the necessary longitudinal cable strength and protection of the relatively fragile core. Charles Bright (son of Charles Tilston Bright) in 1898, described various forms of unsuccessful cables. He wrote:

"There can, however, be no question that, if some form of light cable were devised which, while obviating the various objections - especially that of decay - applying to the original iron and hemp combinations, really possessed the required strength, it would have a great future."

Such a lightweight cable was developed for CANTAT although not primarily for the reasons that Bright had envisaged.

Great care had always been exercised in handling cables at sea in order to prevent the formation of kinks or twists, which can damage or break the cable. The most common cause of kink formation is the untwisting of the armour layer under tension near the ship which in turn results in a twisting-up of the armour wires at the lowest tension point i.e. the seabed. It was this phenomenon which made a deep sea repair in a system with conventional armoured cable and heavy repeater housings almost impossible. The solution was a new cable design that removed the layer of steel armour wires from the outside of the cable and replaced them with a torsionally balanced steel strand located at the centre of the cable. By using high tensile steel wires the cross-section of steel required was greatly reduced and economically filled the centre of the inner conductor. The new cable was called "lightweight" because, in water, it had only about one fifth of the weight of a comparable wire armoured cable having the same transmission performance.

CANTAT went into service on 19th December 1961 providing initially 80 x 4kHz voice channels. While planning CANTAT, agreement was reached between the British, Canadian, American, French and German authorities to re-equip all major submarine cable systems, including TAT-1 and TAT-2 with new high efficiency, 3kHz spaced channel equipment. The development of 3kHz channel technology allowed 33% more channels to be carried over TAT-1, TAT-2 and CANTAT, compared to the original capacities based on 4kHz channels. Further capacity increase was achieved with the Time Assignment Speech Interpolation (TASI) system developed by AT&T, which, by taking advantage of quiet periods in a telephone conversation when nobody spoke, allowed several telephone calls to be sent over the same channel.

Repeaters on-board CS Cable Venture

The Telephone Era 1956 - 1986

Commonwealth Cables

In July 1958, the Commonwealth Telecommunications Conference agreed to recommend to Governments the provision of a complete "round-the-world" telephone system generally to the CANTAT pattern. The estimated cost was about £88 million, including CANTAT and the UK-Canada circuits in TAT-1, which would become the first link in this round-the-world system. This recommendation was accepted in principle by the Commonwealth Governments at the Commonwealth Trade and Economic Conference held in Montreal in September 1958, the UK Government undertaking to provide up to 50% of the total capital. It was left to the countries immediately concerned to initiate action with respect to individual links. In 1959, the Australian Government initiated a Pacific Cable Conference, in Sydney, involving interested parties and recommended the provision of a Commonwealth Pacific Cable (COMPAC) conforming closely to CANTAT. The system linked Sydney to Vancouver via Auckland, Fiji, and Honolulu all being Commonwealth territories with the exception of Honolulu.

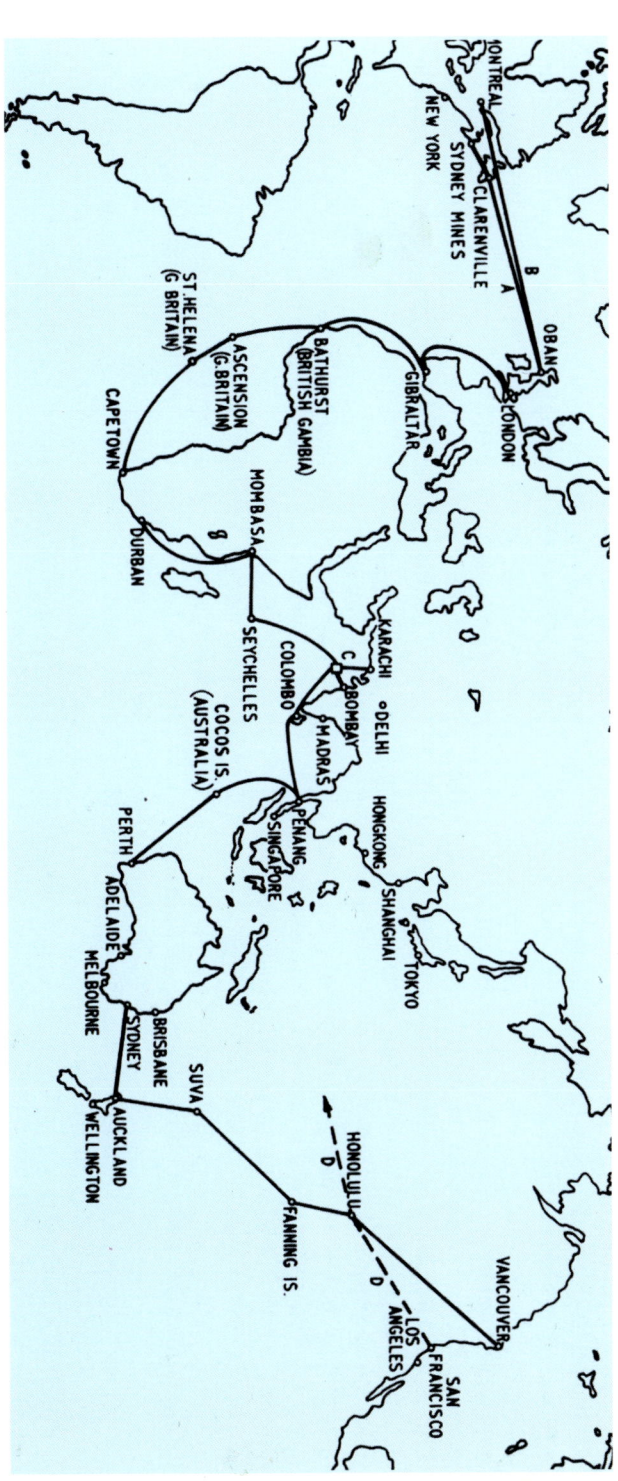

Above
Proposed "round the world" telephone systems

A new cableship was commissioned by Cable & Wireless to assist H M T S *Monarch (IV)* with the installation. She was launched in 1962 and named *Mercury*. COMPAC went into service on 2nd December 1963. AT&T had already connected Honolulu to San Francisco and had announced plans at this stage for telephone cables onward to Japan and other Far Eastern countries. COMPAC connected to CANTAT via the trans-Canada microwave relay system, which was another major link in the round-the-world system. The third link in the chain was called SEACOM; this route connected Malaysia, Singapore, Sabah and Hong Kong, thence via Guam to Madang and Cairns. The Guam landing gave valuable inter-connection with the US to Japan telephone cable. This system went into service on 31st March 1967.

Traffic growth across the Pacific was not as great as in the Atlantic and it wasn't until 1984 that COMPAC was replaced by a new system ANZCAN. This system followed the same route as COMPAC but had an additional landing at Norfolk Island where the Auckland and Sydney systems combined before going north to Fiji. The vast majority of this system had a capacity of 1,380 x 4kHz channels and was provided by STC with the Auckland to Norfolk Island leg manufactured and installed by NEC having 480 x 4kHz channels.

Demand Grows and Competition comes over the Horizon

The technical success of the CANTAT cable and its generation of new voice traffic across the Atlantic gave confidence that further UK-USA traffic could be generated and supported by a new cable, to be known as TAT-3. This system was installed in 1963, and once again required the integration of a new rigid AT&T repeater design with British cable. STC put forward a very strong commercial offer and were awarded the cable contract outright. The system provided 136 x 3kHz channels and continued the tradition of long service, finally being retired in 1986 after 23 years. Again, a system from France to the USA, using similar technology, followed shortly after TAT-3. However, when TAT-4 was installed in 1965, because of the longer route, it could only provide 128 x 3kHz channels.

The year 1965 also saw the first deployment of the Early Bird satellite, which brought the glamour of the space race, contrasting sharply with the "old technology" perception of submarine cables. This perception was translated into investment, and the first set of geostationary Intelsat satellites achieved global coverage in 1968. The use of satellites to carry TV signals immediately struck a chord with the public and even today most people assume that all international telephone traffic is carried by satellite. Even though the delays and echo on satellite telephone circuits were complained about regularly, the relatively large capacities gave network planners the ability to deliver the route and media diversity essential in setting up fully automatic International Direct Dialling services.

Although indirect transatlantic circuits were available in Spain, the need to transit through the existing landing parties in France and the UK led to the demand for a direct route. Significant technological development had occurred to increase cable capacity, and the reduced costs per channel offered by an 845 x 3kHz channel, "high quality" cable, for delivery in 1970, proved overwhelmingly attractive to the Spanish landing parties. This cable, known as TAT-5, offered a strong challenge to the cost of satellite circuits, and managed to reverse the direction

The Telephone Era 1956 - 1986

of transit revenues to the benefit of the Spanish. The system used AT&T repeaters at 18km spacing with 38mm cable supplied by STC and AT&T. It was finally retired after 23 years of service.

Three years later, in 1973, there were strong pressures building for the construction of a very high capacity transatlantic cable from France, to be known as TAT-6. The emerging technology promised more than a thousand voice channels, which would seriously undercut satellite circuit prices even with the greatly reduced repeater spacing required by the higher transmission frequencies. The satellite lobby was very powerful in the USA and this resulted in a three year delay to TAT-6. As a result, the CANTAT-2 system was planned and installed before TAT-6, and went into service in 1974. The entire system was supplied by STC and as a 1,840 x 3kHz channel system it more than doubled the number of submarine cable voice circuits between Europe and North America. CANTAT-2 was in service for 18 years.

The TAT-6 system was finally installed from France to the USA in 1976, and the extra time available for technological development had allowed a capacity of 4,000 x 3kHz channels to be delivered. This was achieved using AT&T repeaters at a 5km spacing, and a mixture of STC and AT&T cables of 53mm diameter (plus armour where required). The cable was twice the diameter of earlier systems, and was correspondingly more difficult and expensive to manufacture. Worse, the short repeater spacing required a total of 693 repeaters in the cable, with the corresponding additional problems of power feeding and signal equalisation.

The capacity of TAT-6 was so large, and took so long to fill, that it was not until 1983 that the matching UK-USA cable, TAT-7, was built. The technology had not moved on at all from 1976, limited as it was by the economics of building such a large co-axial cable with so many repeaters, the 5% increase in capacity can be wholly attributed to the slightly shorter route. The TAT-7 system was retired in 1994.

Although the major high capacity systems were built for the Atlantic route, STC did develop a 45MHz technology, which was used on many European short haul routes. The longest system of this type was PENCAN 3, which went from mainland Spain to the Canary Islands and contained 270 repeaters with 5km spacing. It was the first system to be laid by the CS *Cable Venture*. Installed in 1977, it was designed and equalised to allow 5,520 x 3kHz channels.

Below
Cable & Wireless Cableship Cable Venture

It is clear that there were major problems emerging in the technology and economics of long haul co-axial systems, and that the co-axial cable technology had nowhere to go after TAT-7. One insurmountable problem was the drastic increase in repeater count, which - with such short repeater spacing and a transatlantic water depth of 5,000m - would require a repeater to be overboarded before the previous repeater has reached the ocean floor. Increasing repeater costs also threatened the ability of operators to continue with the cost-cutting strategy that had generated the increased traffic in the first place. As well as the direct cost of repeaters, there were corresponding increases in the number and complexity of the required system equalisers. These smooth out the transmission performance of the system to ensure that all the channels work as well as each other.

Below
Inserting circuit boards in an equaliser

At a time when desktop computing was less easy than it is today, STC used computers onboard ships during system lays in order to design the equalisers, which were inserted into the system after every 15 repeaters. The design was based on measurements made on the partly laid system, which were then adjusted by the computer to forecast performance once all the cable under test had been laid. The equalisers were built in clean room conditions on the ship and inserted into a repeater type housing, already jointed into the cable. This housing then had to be closed and pressurised before laying, without delaying the steady progress of the ship. Other manufacturers used pre-built equalisers and AT&T had a novel system where different pre-built sections could be remotely switched in or out of the line. These methods were quick but never as effective as the STC approach.

A second insurmountable problem relates to the increased size and therefore cost of the cable required. The higher operating frequencies, in the range of 30MHz for TAT-7, required much larger cable if transmission losses were not to become impossibly large, requiring even more repeaters.

The combination of these factors led to a plateau in the previously falling costs of international direct dialled voice circuits. The public still regarded such calls as expensive, and it was clear that a new technology would be required if international calls were to become "business as usual" for network users.

But the key factor that directly triggered the end of the analogue era was the irresistible rise of digital telecommunications. Digital telephony is only possible if you have a method of

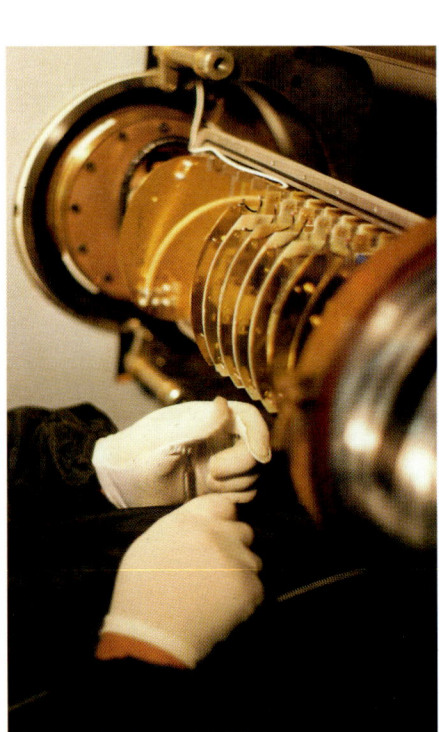

The Telephone Era 1956 - 1986

converting the complex wave pattern of the voice signal into a digital code and back again. This problem was solved in 1938 by Alec Reeves - an STC engineer - when he invented Pulse Code Modulation. PCM was a brilliant technique, but valve technology meant that the equipment was bulky and had high power consumption. It wasn't until the advent of semi-conductors and integrated circuits in particular that PCM became a really practical proposition. From the late 1960s its growth was unstoppable. Analogue cables were simply not able to link the increasingly digital national telephone networks.

The early transatlantic cables had fulfilled their design lives, and were taken out of service when their maintenance costs exceeded their relatively small contributions to revenue stream. It remains as a glowing testament to these early system developers that their systems gave faithful service for over 20 years. The shorter lives of the later analogue systems did not relate to their engineering quality: rather, they all became obsolete in 1992-1994 with the need for a totally digital network. It seems remarkable now that any network element could have remained viable for more than 20 years, but the first analogue technology that was embodied in TAT-1 and TAT-2 was in service right through to the delivery of the last analogue technology embodied in TAT-6 and TAT-7. The latter systems, in turn, survived right through the first decade of optical fibre systems, and were finally retired in 1994 - the same year that Erbium Doped Fibre optical Amplifiers appeared on the UK-Spain route. The last commercial analogue system was manufactured and installed by STC in 1986. It connected India with the United Arab Emirates and had a capacity of 1,840 x 3kHz channels.

Allez la France

The year 1957 marked an important turning point for the Calais factory of Les Câbles de Lyon who had taken over the facility from SIT in 1938. It had just won the contract to provide 1,000NM (nautical miles) of co-axial cable for the second transatlantic cable, TAT-2, which was to connect Penmarch with Newfoundland. The factory, which had been destroyed during the war, was not operational until 1950, during which year the Administration Française des PTT (the French Postal Authorities) laid an experimental telephone cable, which used a repeater equipped with valve amplifiers, between Cannes and Nice. The cable was manufactured at the Bezons and Calais factories and the repeater was made by the Compagnie Industrielle des Téléphones. The French did not follow the American technique: they, like the British, developed bi-directional repeaters. The first commercial application for this repeater was in 1957 with a telephone cable between Marseille and Algiers, which provided 60 channels over one cable. The cable manufactured for TAT-2 was loaded onboard the *Ocean Layer* in June 1959 and the majority of the cable was successfully laid shortly afterwards. Unfortunately, on 14th June the *Ocean Layer* caught fire and was abandoned before the cable lay could be completed. Calais, however, went from strength to strength and produced 1,200NM of lightweight cable for TAT-4 in 1965. The zenith of French achievement in this era was the world's longest submarine telephone system SEA-ME-WE, 11,000km from Marseille to Singapore with intermediate landings in Palermo, Alexandria, over land to Suez then on to Jeddah, Djibouti, Colombo and Medan. The vast majority of this system was manufactured in Calais during 1984 and 1985 and laid by the French ship *Vercors*.

61

Left
Calais factory circa 1950

Cableship Evolution

The advent of rigid repeaters presented cableships with a number of interesting problems. Firstly, the repeater housings could not be stowed in the cable tanks so a method of stacking them on the deck had to be found. Then a way of handling the cable ends, which came out of the tank at regular intervals up to one side of the repeater housing and then from the other side being held down into the tank, had to be devised. Next, in order to send signals through repeaters they need to be powered. For the first time, cableships had to provide high voltage equipment to power the system during the lay, which required very strict power safety procedures to be developed. Repeaters are designed to operate on the seabed at water temperatures of 4°C and therefore when powered onboard ship they need to be cooled by air conditioning, another first for cableships. However, perhaps the most difficult problem was how to lay a long housing which could weigh as much as 750kgm, when it would not pass through the cable-laying machinery.

The development of the new type of rigid repeater had in fact outstripped the development of a satisfactory method of laying it. The first two systems to use these rigid repeaters - Aberdeen-Bergen in 1954 and the Terrenceville-Sydney Mines section of TAT-1 cable in 1956 - had to be laid over the bows with conventional cable gear. The whole operation resulted in the cableship being held without forward movement for some 20-30 minutes thus presenting ideal conditions for the formation of kinks. To solve the problem of laying rigid repeaters in lightweight cable, the Post Office developed a new cable engine, known as the "five-sheave gear" which comprised five v-grooved sheaves in tandem around which the cable took less than a complete turn. The inner three sheaves were directly coupled by means of a roller chain and large diameter sprockets. The repeater was mounted on a trolley and bridged by a flexible steel rope that was spliced to the cable approximately four fathoms (1 fathom = 6ft or 1.83m) ahead of and behind the repeater. This rope was termed a "by-pass rope" and was made shorter than the cable and repeater it by-passed. Shortly after the leading splice entered the cable engine the cable was guided to one side and the rope took the place of the cable in the sheaves. The tension was smoothly transferred from the cable to the by-pass rope and vice versa with the trailing splice.

The sheaves were constructed as continuous v-grooves in the initial design. However, lightweight cable presented a potential problem because the strength member of the cable was in the middle and the laying machinery gripped the outside. In circumstances where similar coefficients of limiting friction exist between the various components of a cable and there is no bond or adhesion between them, slip will occur at the interface having the smallest diameter, i.e. between the strength member and the inner conductor. In other words there was a very good chance the cable would pull itself apart when it was laid. In practice, the so called adhesion forces are automatically built into the cable during manufacture and the minimum values were always specified for the supplier. However, it was difficult to ensure that the required adhesion would be achieved during large-scale production. As a precautionary measure, steps were taken to develop a resilient or shear-limiting sheave, to replace the three coupled sheaves of the five-sheave gear. The periphery of the replacement sheaves was divided into 24 segments, each linked to the boss of the sheave by two pairs of radial flat steel springs. The use of two sets of springs ensured that each segment moved in a line that approximated to the arc of the cable on the sheave and the springs in each of the three sheaves were also pre-formed to provide different pre-load pressures. In operation, as soon as the load due to cable tension on any one segment reached the pre-load value, the segment tended to leave the backstop and increased the load on the next segment. The overall effect being to make the rate of decay of tension in the cable around the sheave linear.

The five-sheave engines were successful for several years but they still had the major drawback that the ship had to slow virtually to a standstill, albeit for a very short period. To counter this AT&T developed a caterpillar engine called a "Linear Cable Engine", which was installed on its flag ship *Long Lines* in 1963. The engine was very similar to a caterpillar engine developed for the *Ocean Layer* earlier. This engine had been a failure because the tracks became clogged with tar. With the advent of lightweight cable this became less of a problem and it is interesting to note that for the majority of her career *Long Lines* only laid lightweight cable. The most successful linear cable engine was developed by the British Post Office and Dowty Boulton Paul Ltd. It consists of a number of rubber tyres in pairs that press vertically up and down on the cable and move apart to allow the repeater to pass through. The first engine was installed on the *Alert (IV)* in 1971 and comprised 14 individual wheel pairs. By 1973 a vastly improved engine was installed on the Cable & Wireless ship *Mercury*. It had 18-wheel pairs grouped in threes, and this design has now become the basis for an industry standard.

Right
Five-Sheave cable engine

Below
The original linear cable engine on Cableship Alert

The Telephone Era 1956 - 1986

The sinking velocity of the lightweight cable was also much lower than the rigid repeater and this caused great concern. The problem was solved by fitting parachutes to each repeater to equalise the sinking velocities. They were made of silk or viscose rayon, 28ft in diameter and were successfully tested in Loch Fyne in October 1960. They were fitted to the repeaters with a hydrostatic device that uncoupled the parachute after use - to reduce recovery tensions in the event of repair. As confidence grew and cables got larger in diameter these parachutes were phased out.

Although satellite communication was in competition with submarine systems, it was a positive boon to cableships in two key areas. In the early 1970s the advent of "Transit" satellites provided a major step forward in the accuracy of navigation and therefore the placement of the cable, enabling the percentage of system slack (excess cable over route length) to be reduced. Until then slack had been managed by the use of Taut Wire, a system introduced by William Siemens in 1874 on his cableship *Faraday*. It measured the distance the ship sailed by paying out a thin wire in parallel with the cable and comparing its length with the length of cable laid. As their name suggests, Transit satellites worked by measuring the Doppler Effect as the satellite passed overhead. At the time of a satellite fix, the accuracy was very high, but coverage was limited, and in many areas of the world several hours could pass between fixes. In the 1980s a Geostationary Positioning System was launched, initially for military use. This gave global coverage and constant updates on the ship's position. The absolute accuracy of this system was comfortably within the length of the ship, anywhere in the world. Geostationary satellites, through the INMARSAT system, also improved ship to shore communication, an essential part of cable-laying operations. Direct telephone access to the terminal station, away from which the cable was being laid, significantly improved the efficiency of powering and de-powering the system. Unfortunately, this enhanced communication capability also increased the demands on the cableships to produce daily reports. Where in the past a three-line telex had been sufficient a 20-page faxed report became the norm!

Finally, as we have already seen, repeatered systems needed equalisation, requiring the ships to set up factory-type clean room environments. In addition, as the transmission loss of co-axial cable is temperature dependent, it was important for the temperature spread across the cable in the tanks to be as small as possible. This was achieved by inserting thermometers into the flakes of cable during loading and spraying the cable with fresh water.

These modifications to ships required significant capital investment, most of which was provided by the major cable owners and, by 1976, all cable-laying and maintenance ships were owned and operated by them. Once again the dominant forces were the British, Americans and French. By far the biggest fleet was operated by Cable & Wireless. It had one laying ship and maintenance ships located in Bermuda, Fiji, Hawaii, Singapore and Vigo in Spain. In 1983, Cable & Wireless set its marine division up as a wholly owned subsidiary, Cable & Wireless (Marine) Ltd. British Telecom also operated a fleet of three ships out of its Southampton depot which opened in 1974. The American ships were owned by AT&T and the French fleet by France Câbles et Radio, a subsidiary of France Telecom.

Monopolies and the "Big Four"

The Telephone Era is quite unique in one respect. The cable network in the Telegraph Era had been built largely on private investment and competition was available on most routes. However, with the nationalisation of Cable & Wireless in 1947 the vast majority of international telephone traffic came under the control of government owned utilities with a single monopoly operator in each country. We, the customers, became *subscribers* with no choice other than to take it or leave it. This situation had a significant effect on the manufacturing industry. Government-owned companies when investing in new systems would press hard for national content. Therefore, the major investors in submarine cables had the greatest leverage or influence in directing contracts. This resulted in the manufacturing industry becoming dominated by four main players. In Britain, the Post Office and Cable & Wireless supported Submarine Cables Ltd (originally Telcon) and STC. However, by 1970, STC's better repeater technology and its modern cable-manufacturing capability, combined with aggressive marketing, allowed it to take over SCL and become the sole British supplier. In the same year the French industry had consolidated under Alcatel Submarcom, who were supported by France Telecom. In the USA, AT&T was the monopoly carrier and moved towards vertical integration through contracts with Bell Labs, Western Electric and the Simplex cable factory. Finally, they set up their own subsidiary AT&T SSI. During this period the Japanese industry came of age. Its cable was made by the Ocean Cable Company and submerged electronics were made by Fujitsu and NEC. These companies shared the work between them, never bidding against each other, they became known in the business as Japan Inc. Japan Inc. was strongly supported by Kokusai Denshin Denwa the Japanese international carrier. As the Telephone Era drew to a close the world began to turn away from government monopolies and back to private enterprise. With this, came a significant decline in protected markets. With a new technology on the horizon and reduced levels of patronage, the 'Big Four' were preparing to do battle on a vastly changed playing field.

Light at the End of the Tunnel

The stagnation of co-axial cable technology that arose with TAT-6 in 1976 coincided with the first glimmerings of the possibilities of optical fibre. Although digital co-axial systems had been considered, they offered no benefit in terms of reduced cable size or increased repeater spacing problems. Considerable efforts had already been made to develop waveguides for high capacity terrestrial use in national core networks, but all that work was dropped overnight in 1976 when the potential of optical transmission was first appreciated. At that point the British Post Office had achieved losses of less than 3dB per kilometre from 30 to 110GHz, with less than 0.1dB loss in realistic bends, which was truly remarkable - but the 50mm diameter hollow waveguide structure was completely impractical for deep-sea use.

The roots of optical fibre can be traced back to a decade earlier, to 1966 with the publication of Kao and Hockham's famous paper, and a visit of a Corning scientist to Post Office engineers in London to discuss the application of optical fibre for telecommunications. Spurred on by the competitive development from European laboratories, Corning pulled out all the stops to achieve a result of 16dB per kilometre in 1970,

The Telephone Era 1956 - 1986

which was about a thousand times lower than common glass. Further work produced 4dB per km and 4km lengths in 1975. Corning was active in generating patents and this came to be a crucial advantage for it later on. The first practical semi-conductor laser was achieved by Bell Labs in 1970, and, by using a fibre made by a CVD (Chemical Vapour Deposition) process in conjunction with their own detector, they were able to carry out a field trial at Atlanta, Georgia in 1975. This ambitious trial used a cable operating at 820nm (nanometre) wavelength and at 45Mbit/s. Parallel work by the British Post Office engineers - Dr Marc Faktor, Dr George Newns and Dr Keith Beales - had led to the successful development of the "double crucible" fibre-manufacturing process and, perhaps more significantly, the early understanding of the importance of single-mode fibre. These were the building blocks, which were to herald the dawn of a totally new era for submarine cables.

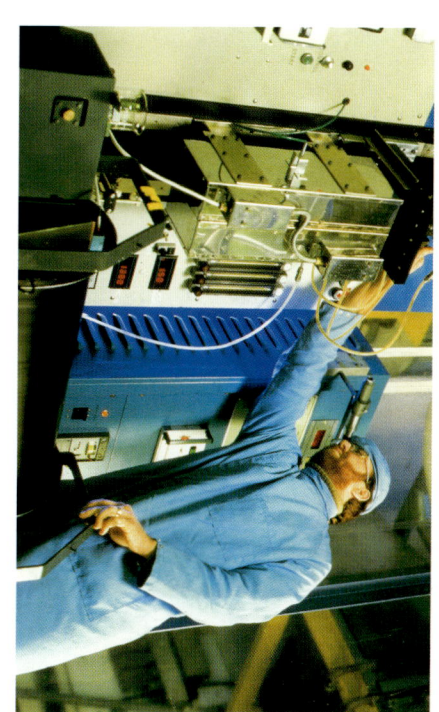

Above
Optical fibre pulling tower, STC Harlow, UK

"The Light Fantastic"

Section 4 The Optical Era 1986 - 2000

Just a few years ago a phone call from Europe to New York or Tokyo was an expensive proposition, even for companies. Now you hardly have to think twice about it. Today, the cost of a transatlantic call is a matter of pennies rather than pounds, thanks to two closely linked phenomena: the deregulation of telecommunications which has unleashed ferocious competition between the major players in Japan, France, the USA and the UK and the deployment of intercontinental submarine fibre optic cables. These two events have revolutionised the industry and, just like the laying of the first submarine cable in 1850, are changing the world - at warp speed.

Revolutions, of course, don't just happen. Twenty years before the first commercial submarine fibre optic cable was laid in 1986, two British scientists working at Standard Telecommunications Laboratories (STL), the research division of STC/ITT in Harlow in the UK, Dr Charles Kao and Dr George Hockham, reported a major discovery:

> *"A fibre of glassy material constructed in a cladded structure with a core diameter of about λ_0 and an overall diameter of about $100\lambda_0$ represents a practical optical waveguide with important potential as a new form of communication medium...compared with existing co-axial cable and radio systems, this form of waveguide has a larger information capacity and possible advantages in basic material cost."*

PROC. IEE, Vol. 113, No. 7, July 1966

"Larger information capacity" proved to be something of an understatement. When finally glass fibres replaced traditional copper cables the amount of traffic that could be squeezed into a single strand of cable leapt from 5,500 to 12,000 channels

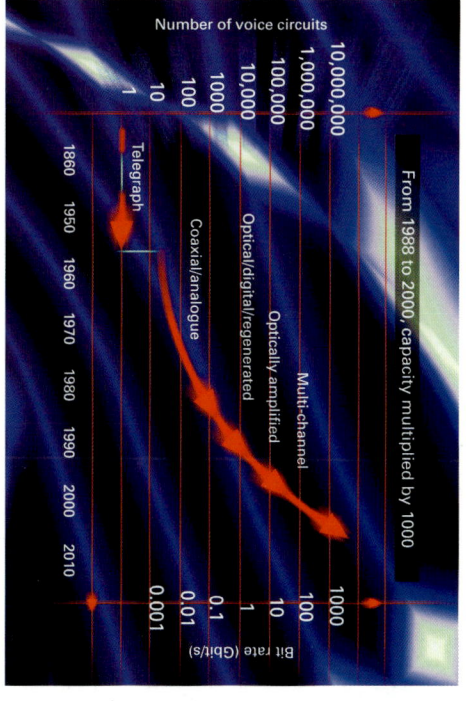

Left
Optical fibres

Above
Evolution of technology and capacity

but this was only a foretaste of what was to happen in the following years. What nobody realised in the late 1980s was that the ribbons of light beginning to gird the globe wouldn't just revolutionise conventional telecommunications. Optical cables were also wiring the planet for the Internet. The first submarine cable back in 1850 carried a single channel: today each fibre has a capacity of a mind-boggling 15 million channels. Without them the Internet would soon hit gridlock.

Pure Genius

In 1966, Kao and Hockham had pointed out that the attenuation of glass fibres was not a fundamental property of the fibre itself but was caused by impurities. Reduce the impurities sufficiently and an attenuation of only a few decibels per kilometre, or even less should be achievable. The significance of their proposal was widely realised and led to considerable research effort in the UK and also in the USA, France, Japan and Germany.

Over the next decade scientists continued to refine the technology along with the chemicals used in the production of the optical fibre, which required purity levels higher than 99.9999%. In 1977, over a 4km stretch between Hitchin and Stevenage, north of London, STL installed the first experimental high-capacity optical system in the UK. The system operated at 140Mbit/s and at a wavelength of 850nm transmitted over graded-index multi-mode fibre and, significantly, proved itself competitive with co-axial systems. But we were still on dry land: it was high time fibre optics got its feet wet.

Right
Charlie Kao

Below right
George Hockham

The Optical Era 1986 - 2000

Cables and Kippers

In 1980 STC, working with British Telecom, installed a short experimental submarine system in Loch Fyne, a seawater loch located to the north-west of Glasgow hitherto renowned in Britain for the excellence of its kippered herrings and the beauty of its scenery.

The cable for the Loch Fyne project, the world's first trial of an optical fibre submarine cable system, was manufactured in late 1979 and laid by the British Telecom vessel *CS Iris (III)* (a good name for an optical cableship) the following February. Six fibres, four of which were multi-mode for operation at 850nm, and two single-mode test fibres were packaged into the 10km cable which was laid in a loop at an average depth of 100 metres with both ends installed into the nearby Inveraray repeater station. The cable was subsequently recovered and a submerged repeater spliced into the system before being re-laid.

These experiments demonstrated that the system had the necessary stability of transmission characteristics and was able to cope well with the mechanical stresses involved in getting the cable off the ship, onto the loch floor and later recovery for repair.

A Marine Milestone

Following the trial in Scotland, all of the world's leading manufacturers, including NTT and KDD of Japan, Alcatel of France and AT&T of the USA, launched major development programmes of their own for submarine systems. The fight to come up with the best technology was on.

Above
The World's first optical system being laid in Loch Fyne, 1980

Above
Repeater loading for Mainland Corsica

The experience in Loch Fyne led to the design of the cable that was used in UK-Belgium No. 5, which was the big commercial breakthrough. The UK-Belgium project was the world's first international underwater optical fibre telecommunications system and operated at the longer wavelength of 1,310nm with single-mode fibres. The work was completed on 4th May 1986 and the link was opened for commercial traffic shortly afterwards. Fibre optics was finally beginning to deliver its promises.

The cable contained six single-mode fibres and could carry a total of 12,000 simultaneous two-way telephone conversations. In a fibre optic system, there are three windows for transmission, 820, 1,310 and 1,550nm. At the longer wavelengths the loss in the fibre is less, this allows greater distances between repeaters and, fewer repeaters means lower costs. The first system crossed the English Channel over a distance of 112km and contained three repeaters.

At the same time, AT&T and Alcatel, working with France Telecom developed their first commercial systems, Optican and Mainland - Corsica respectively. Both were deployed in deep water in order to experiment with new lightweight cables.

It soon became clear that fibre-optic technology offered the prospect of some even more radical technical and commercial solutions. The 133km UK-Channel Islands No. 7 link, operating at the 1,550nm wavelength and completed in 1988, was the longest commercial "repeaterless" system in the world at the time. The system used single-mode technology, high-performance lasers and a new Avalanche Photo Diode or APD receiver, that enable the light signals to travel much further before they need to be regenerated, thus requiring fewer repeaters or, in this case, none at all. The six-fibre pair system had the same capacity as the UK-Belgium No. 5 link with the added advantage that the terminal equipment could be upgraded to operate at 2.5Gbit/s at a later date. The only downside with repeaterless submarine optical cable systems is that you are limited, even today, to a maximum length of 300-400km. So you can't, with current technology, span intercontinental distances without using repeaters.

Crossing the Pond

Up until the 1960s, the North Atlantic was the most prestigious route for luxury ocean liners and has always been the Blue Ribbon challenge for transoceanic submarine cable technology. Once the new optical technology had been put through its paces over relatively short distances, planning got underway on this Atlantic challenge. AT&T was the lead contractor for the TAT-8 system. The first intercontinental system anywhere to employ optical fibre digital technology. TAT-8 comprised two pairs of fibres. One pair between the USA and the UK, the other between the USA and France, the two pairs separating at a branching unit (located off the European Continental Shelf). A separate pair of fibres connected France and the UK, giving four fibres in each leg of the system.

The advent of branching units brought a new system component, which allowed countries, often smaller nations, along the way to take advantage of the network being laid by dropping a spur cable rather than installing a point-to-point link. The first system of this type was TAT-8, using technology developed by AT&T.

The Optical Era 1986 - 2000

Fittingly, the whole pioneering venture optically linking Europe and North America was an international collaborative effort involving three leading suppliers: AT&T of the USA, STC of the UK and Alcatel Submarcom of France. Their three different designs presented the greatest wet integration challenge the industry had attempted. Naturally, these three manufacturers had different philosophies in their approach to achieving the specified system reliability. The TAT-8 contract required a design life of 25 years and only three ship repairs due to equipment failure; ship repairs take a lot of time and cost a lot of money. Each of the suppliers set up a comprehensive life-testing programme. They also used redundancy arrangements to a greater or lesser extent. Finally, test programmes were initiated to demonstrate that the three designs, along with the three different prototype repeaters, were compatible and worked together successfully. They did and TAT-8 was commissioned on schedule in October 1988. It operated at a relatively modest 280Mbit/s and with a capacity of just under 8,000 x 64 Kbit/s channels (64 Kbit/s digital channel is equivalent to a 4kHz voice channel).

Private Investment Returns

By the early 1980s, AT&T began to assume a monopolistic position in the market and America's Federal Communications Commission pounced. On 1st January 1984 the FCC broke up AT&T, giving birth to half a dozen operating companies called "Baby Bells" in the process. The UK followed shortly after when the Conservative Government, under Margaret Thatcher, returned Cable & Wireless to private investors. This kick-started the world-wide deregulation of the telecommunications industry. Next came the privatisation of BT in UK followed in due course

Above
TAT-8 branching unit

by the gradual and often less than enthusiastic liberalisation of many other state telecom monopolies across mainland Europe. Due largely to this deregulation within the telecoms industry, the optical era for submarine cables got underway with a vengeance, with tenders issued for projects like PTAT-1 and TPC-3 required in 1989, EMOS in 1990, and TAT-9 and NPC in 1991.

Now the competition was between companies rather than between countries and it was hotting up. EMOS, linking Italy, Greece, Turkey and Israel, marked the beginning of the era of "gloves-off" competition between the leading suppliers. Every system was a little different, influenced by the different technological approaches of the now fiercely competing manufacturers of the four main producer countries, France, Japan, the UK and the USA. With the unfettered interplay of market forces now operating, new suppliers, often from other countries, were able to enter the market. Siemens of Germany developed an innovative cable design for repeaterless systems using a miniature basic cable dubbed "Minisub"; the Italian company Pirelli laid a festoon around Italy; STK of Norway, Ericsson of Sweden and NKT of Denmark all developed their own repeaterless technologies.

A freed-up market also meant survival of the fittest. On the EMOS project, Alcatel with their fibres cocooned inside a steel protective core were able to demonstrate a key advantage over STC to secure the contract. Following its success on EMOS, in 1992 Alcatel was awarded another major project, TASMAN-2, linking Australia and New Zealand and set up a cable manufacturing plant near Botany Bay. In late 1988, STC and NEC working in consortium were awarded the NPC (North Pacific Cable) contract and the following year STC set up a new

Right
Alcatel cable factory
Port Botany, Australia

Below right
Alcatel cable factory,
Portland, USA

cable factory in Portland, Oregon. Both Alcatel and STC were going global while thinking local, and the upshot was that by March 1994 Alcatel had purchased STC from its owners, Northern Telecom to form Alcatel Submarine Networks. The unification of Alcatel and STC's research and development activities, together with the integration of their manufacturing capabilities and resources, was a key milestone in the development of the industry.

Advancing further around the Globe

In mid -1988, a link delivering similar capacity to TAT-8, around 8,000 channels, was installed between the UK and Denmark. The difference was that for the first time the repeaters were spliced into the cable in the factory, rather than on the ship during loading. This had the double advantage that the system could be tested for overall integrity in comfort on dry land and then be loaded onboard the ship in a much shorter time. The cable industry was hacking its way up a steep learning curve but it was getting there.

In 1989, STC increased cable capacity by the introduction of 420Mbit/s technology and this was employed, that same year, in the first privately sponsored transatlantic cable, PTAT-1. This was the first complete transatlantic fibre optic cable to be supplied by a single contractor, with a capacity of 18,000 channels. The cable was notable because it had more service channels (channels used by the system for housekeeping) than the total capacity of the original TAT-1 system.

Two years later TAT-9, the first repeatered long wavelength (1,550nm) system, and the first to operate at 560Mbit/s, was installed. An underwater branching multiplexer was used at both ends. The European end was divided into spurs to France, Spain and the UK with the multiplexer splitting the traffic. The USA end was split between America and Canada. This system, along with TPC-4 (Trans Pacific Cable 4) in 1992, heralded the beginning of a new generation of 1,550nm regenerated optical systems with significantly greater repeater spacing.

By now fibre optic cables were beginning to snake out from land mass to land mass across the planet. These included Denmark-Russia in February 1992, Russia-Japan-Korea in November 1994 and SEA-ME-WE-2 in mid-1994, which connects Europe to the Far East with a complex multi-landing network. In August of the same year, the pace-setting CELTIC (UK-Ireland) system, was commissioned. It was an impressive 270 kilometres of unrepeatered cable, operating at 2.5Gbit/s.

In the early 1990s, based on the 1985 work of Drs D Payne and S B Poole at Southampton University, the industry began to develop optically amplified systems to achieve longer repeater spacing for a specific bit rate. They were used initially for land cables, and then in repeaters for submarine cable, compensating for the fibre loss without the need for any optical/electrical conversion of the signal. The first fully optical systems, which were laid in 1995 and 1996, were TAT-12/13; both systems operating at 5Gbit/s. Two cables rather than one were deployed across the Atlantic in a ring configuration with one cable protecting the other. Because of the very high capacity of each cable, it was seen that if a single cable was damaged it would be a disaster for the overall transatlantic telephone business. The other early, amplified system was the 2.5Gbit/s Rioja system linking Spain-UK-Belgium-The Netherlands. This was the first system to deploy a ROPA or Remote Optically Pumped Amplifier.

Wave upon Wave

About the time TAT-12/13 was installed, the scientists suddenly produced another couple of tricks from up their sleeves. Firstly, they showed that it was possible to double capacity by adding a second wavelength and that this principle could be extended by up to as many as 16 wavelengths. This technology was branded Wavelength Division Multiplexing or WDM. Secondly, a technique that Alcatel had been working on for some years, Forward Error Correction became commercially viable. This allows the spacing between repeaters to be increased further and has become an essential element of all modern submarine systems. FEC combined with WDM allowed a three-fold increase in the traffic carrying capability of TAT-12/13. This was over-capacity with a vengeance and a shock for both suppliers and operators who thought they suddenly had too much capacity on their hands, and not enough traffic. But this was back in 1995 when the full impact of the Internet had yet to sink in. The relentless explosive acceleration in capacity continued further with the development of DWDM technology, Dense Wavelength Division Multiplexing, which was soon capable of delivering up to 640Gbit/s per fibre pair (64 wavelengths at 10Gbit/s). DWDM takes advantage of the wide bandwidth of line fibre and optical amplifiers by combining multiple traffic streams into a single fibre. Each traffic stream has a different wavelength (colour) and can be combined and separated using optical components. The technique is analogous to splitting white light into the colours of the rainbow using a prism.

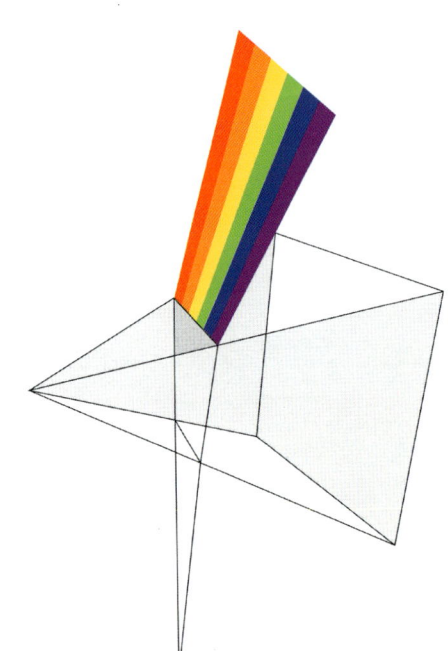

Below
Prism splitting white light into its constituent colours

Far right
A pioneering optical cable plough

Large Sharks, Subatomic Sparks and Burial at Sea

You might think fibre-optic cable laid underwater is relatively safe from damage but in fact the hazards are legion. Sharks, for example, have big jaws, sharp teeth and sometimes peculiar feeding habits.

AT&T was the only supplier to report shark damage, which happened on their early experimental Optican system in the Canary Islands. The cable appears to have been suspended above the seabed, it was therefore moving in the currents and at the same time creating an electrical field. A shark was attracted and took something of a mega bite. The fish lost a tooth, which became imbedded in the cable, so it's unlikely it would repeat the experience. Unfortunately it also caused a fault to the cable. The industry's answer was to develop and use lightweight screened cable, which has an outer metallic screen and has been successfully deployed around the world in areas of perceived high shark-bite risk. The screening makes the cable more like the co-axial lightweight of the telephone era and eliminates the minute electromagnetic effects that sharks appear able to detect and which apparently make the fish think they've found a tasty meal. Beyond 2,000 metres water depth this screen is not required because no sharks are found at these greater depths.

Over its projected lifetime of 25 years, a submarine cable has a lot to put up with. It has to withstand storage, laying, plough burial in water depths of up to 1,500 metres, potential de-trenching and then hauling up to a ship for repair as well as being jointed in a simple and speedy manner. It also has to withstand the perils of the sea. Not just mean fish but also

deep-sea pressure in the ocean depths, abrasion in shallower water, mile-wide trawler nets and ships' anchors.

In the early 1980s, with trawlers increasing in size and with optical cable on the horizon, British Telecom International, after initially opting for bigger and better armour, decided that burial provided greater and less expensive protection. This meant a reappraisal of cable designs and a commitment to developing new cable-laying and recovery techniques.

The first question was how deep should the cable be buried, to best protect it from anchor or trawl damage? After studying the literature and analysing fault reports, a good compromise was struck between security and economics by specifying a burial depth of 600 mm, safe from conventional fishing techniques in most soils but limited protection against anchors.

The next big issue was the methodology. British Telecom International, in collaboration with Soil Machine Dynamics, developed a unique design for a ship-towed plough which underwent a successful sea trial in early summer 1986 and was immediately put to work burying an 88km stretch on the ground-breaking UK-Belgium No. 5 system. This new plough could plough deeper, left lower residual tension in the cable and back filled the trench after minimal disturbance of the seabed. Two years later the same plough was used to bury the UK section of TAT-8 and the UK-Denmark No. 4 cable systems. This plough was a major step forward from the "sea plow" developed by AT&T to bury co-axial cables and has become the de-facto industry standard.

The new plough's performance was exemplary but it wasn't really suitable for burying the final splice of a submarine system nor was it economical for reburying short lengths of cable after repair. In the 1970s, France Telecom in collaboration with the French firm SIMEC, successfully developed a tracked trencher called Castor (French for beaver), which was used to bury cable on France-UK No. 4 and UK-Belgium No. 6. It was a bottom crawling remotely controlled tractor, ideally suited to the strong tides and currents found in the North Sea and English Channel. British Telecom designed a similar unmanned submersible known as the BTI Trencher. The Trencher was designed not only to uncover buried cable for fault location and recovery but also to jet a trench alongside surface-laid cable into which the cable could then be lowered. All by remote control from the cableship. This vehicle still operates in the North Sea as a maintenance tool for remedial burial. A third type of burial tool is the free swimming ROV (remotely operated vehicle). This is attached to a mother ship by an umbilical and can carry out fault location, burial, de-burial, and recovery. In response to the increasing depth of commercial fishing these vehicles can operate down to 2,500 metres water depth, well beyond the range of ploughs or tractors.

This improved burial technology achieves three highly desirable goals: it reduces armouring costs substantially; it greatly enhances the reputation of submarine cable for reliability; and it means that trawlermen have one less headache to contend with.

Cableships - the Story Goes On

The development of optical systems once again changed the requirements for cableships. Power feed currents increased significantly bringing with it a number of problems. More current meant that repeaters generated more heat and so repeater stack-cooling systems needed to be enhanced. The development of lightweight cable without a screen meant, for the first time it was unsafe for people to be in the cable tank when cable was being deployed, if the system was powered. Coils of this lightweight cable have massive inductance, so surge protection is vital when powering or de-powering the system. The electromagnetic field generated by this inductance when the system is powered is, in some cases, sufficient to deflect a ship's gyro-compass by as much as 40 degrees.

The weight of optical cable has increased and its diameter reduced compared to co-axial cables, which presents a new set of handling problems. In many respects, in terms of its handling characteristics, optical cable behaves more like its telegraph ancestor than its co-axial predecessor, so many old

Cable Innovator, a modern stern working cableship

skills had to be re-learnt. Deploying and recovering branching units, however, was a new problem and it initially proved to be a challenging proposition, due to the fact that the ship had to deal with three cable ends all at once. Also, the black art of slack control, for deep-water cable-laying, has given way to computer-aided tools such as CASCADE or ESPADON. Now with the processing power of modern computers, the deployment of cable can be simulated and tested in the office using the actual manufactured cable lengths and the survey data. A slack plan can be adjusted where the risks of cable suspensions (not enough slack) or loops (too much slack) are identified. The detailed plan can then be finalised before the ship puts to sea.

The off-shore oil industry had developed powerful propulsion systems to allow vessels to operate close to oil rigs. These differential positioning or DP systems as they are known, are ideal for the slow accurate manoeuvres required for cable work. The cableship operators quickly embraced these developments, firstly by converting off-shore vessels for cable work and later by building new ships including DP technology. For as long as anyone can remember, cable has been laid over the stern of the vessel but when cable had to be recovered or final splices slipped it was always done over the bows. This was entirely due to the position-keeping capability of the ships. When a ship is standing over a cable end, the cable must not be dragged around the sea floor or it will become damaged. With the early ships the only way to achieve this was to put the ship's head up into the wind and keep the bow as still as possible by manoeuvring the stern around it with the ship's engines. The advent of DP solved this problem. The system allows the stern just as easily as the bow to be held in a fixed position relative to the seabed. This has removed the need for the tricky operations of transferring cable from stern to bow and from bow to stern during cable operations. It also has the added advantage that work on the stern cable deck is carried out in the lee of the vessel's superstructure as opposed to in the teeth of a gale on the foredeck. Today, modern cableships are almost exclusively stern-working only.

The increased commercial pressures of the optical era have also had a significant impact on the ownership of cableships. On 1st October 1987, British Telecom followed Cable & Wireless's lead by making BT Marine a wholly owned subsidiary company, and in 1995 BT Marine and Cable & Wireless (Marine) merged. Shortly afterwards AT&T sold its fleet along with its submarine system manufacturing arm to Tyco Submarine Systems Ltd (TSSL). In 1999, TSSL also purchased Temasa, the cableship arm of Telefonica of Spain. In July 1999, Cable & Wireless Marine was purchased by Global Crossing Ltd. becoming Global Marine Systems Ltd. and in January 2000 France Telecom formed its cableship division into a wholly owned subsidiary FT Marine.

Cable Design

From the late 1980s, cable burial was adopted more and more frequently. This option meant that the cable designer had to determine the various de-trenching forces this could entail and to take them into account in the design of the finished product. The cable had to be rugged enough to cope with being buried and then possibly drawn up to the surface for repair work. But it required less armouring than cable laid directly onto the seabed. To make sure the cable delivers the required performance at a realistic cost, fibre-optic cable designers have

The Optical Era 1986 - 2000

Above
An optical cable family showing double armoured, single armoured and lightweight designs

Getting into Deeper Water

Since it is desirable that all the cables in the system be compatible, the ideal solution is to have the same basic cable with the same central package throughout, with outer layers of protection added for those zones that require it. In other words, varying amounts of armour are needed to cater for the range of hazards to cables, which range from, damage and snagging by trawl nets and anchors to natural hazards such as rock abrasion or slumping of the seabed. There are also a host of potential perils when the cable emerges onto the beach and finally goes overland where it is routed via ducts to the terminal building to connect with the terrestrial network.

Cables laid in deep water can be unarmoured or lightweight since the deep ocean floor is generally a benign environment where they are unlikely to be exposed to damage from trawlers and ships' anchors, although the cable still has to withstand the immense pressures that exist in the ocean depths as well as occasional strong currents and mountainous seabed profiles.

Another factor is that the steel strength members in optical fibre cables are susceptible to damage by seawater, if exposed. If the cable is accidentally severed, water can penetrate quite a long way along the cable. The challenge was to come up with a material that could be injected into all the interstices without compromising the properties of the cable or the fibre but that would effectively stop water ingressing great distances along the cable while waiting for the arrival of the repair ship. Modern cables are fully waterblocked with highly effective materials that prevent this problem.

Another wrinkle cable designers have to deal with is the hydrogen generation of the cable components. This is because glass fibre is particularly sensitive to increased loss caused by the presence of hydrogen. So it is important to take into account the extent to which the cable components generate hydrogen both individually and in interaction with each other and to limit hydrogen ingress into the cable from any other source.

It is in fact to optimise a whole range of parameters - mechanical, electrical and optical. Among the factors that have to be taken into consideration are protection against deep-sea water pressure, conductor resistance, electrical stress, water ingress, hydrogen susceptibility, tensile/torsional strength, and resistance to wear and fatigue. No easy task!

This means the cable designer has to review as many materials as possible singly and in combination to find the optimum components for the cable structure.

The Problem of Dispersion

As system bit rates increased a new problem appeared. The problem was dispersion or spreading of the light pulse. It was recognised that these new types of submarine optical fibre systems were required for two main types of link. The first are high bit rate (Gigabit to Terabit) transoceanic links with repeater spacings of 50 to 100km. The second are high bit rate systems (2.5Gbit/s and 10Gbit/s per wavelength) repeaterless spans up to 430km between islands, from the mainland to an island, a coastal festoon or a close coast link. In both areas the effects of dispersion were becoming increasingly significant and had to be dealt with.

Initial attempts to develop so-called "dispersion-shifted" fibre resulted in an unacceptable increase in attenuation or loss within the fibre. It seemed that either low attenuation or low dispersion was possible but not both at the same time. The conundrum was initially solved by adjustment of the optical fibre core profile. The resulting dispersion-shifted fibre was used successfully, right at the very tail end of the twentieth century, on Rioja, TAT-12/13, TPC-5 and on Gemini, the first transatlantic single supplier optically amplified system. Now, the problem for DWDM systems is not dispersion itself but the dispersion slope of the fibre between different wavelengths. To resolve this issue, a number of new fibre designs are being introduced and system design is being further complicated by the fibre span between two repeaters comprising more than one fibre type. This approach to dispersion management requires owners to maintain detailed fibre maps of the system and store a wider range of spare cable for repair operations. An interesting new challenge for the owner, manufacturer and marine maintenance provider!

Joining Hands Across the Sea - The Universal Joint

For TAT-8, the three different suppliers came up with three different cable designs. Each design provided a home for the fibres along the neutral axis of the cable. Features such as, tensile strength, protection from water pressure, hydrogen and water ingress were provided by water blocking and other concentric layers of the cable. Hydrogen was a major concern during the design of these optical cables once its negative effects had been established. A layer of insulation was provided to protect the high-voltage, power-feeding current on the central package. In shallow water, the cable was further protected by one or more layers of armour wires of varying sizes according to how much protection was required. Although they appeared to have common problems to solve, each supplier adopted a significantly different design philosophy, which prompted a different in-house approach to jointing and jointing equipment. The owners of TAT-8 found themselves in the position of having to buy three jointing technologies to maintain their system. This was clearly an expensive and unsatisfactory situation. Therefore, when TAT-9 was being planned the owners agreed to fund the development of a single technology to join all the cable designs. This resulted in the Universal Jointing Consortium being formed in 1989 to provide the industry with a single jointing technology for system maintenance. The original members of the consortium were Alcatel, AT&T and BT Marine

The Optical Era 1986 - 2000

Above
A cutaway of a Universal cable joint

Cable systems used to be owned by a small group of big telephone companies such as AT&T, France Telecom or British Telecom. Nowadays it could be anybody. For many years the floating of new transoceanic cables was achieved by the forming of large consortia known familiarly as "clubs". This was the ownership structure adopted for APCN, TAT-12/13, TPC-5, China-USA, Japan-USA, TAT-14, and SEA-ME-WE-3 and many others before and indeed since. These clubs issue a detailed specification and adjudicate a lengthy, competitive tendering process. However, the cost of running such consortia and the bureaucratic inertia they can produce has resulted in new start-up companies acting more nimbly and being first to market by adopting different philosophies. Being first to market is critical where the system operator is selling the capacity externally rather than using it for its own needs.

One of these different approaches is the sponsored cable network where a single customer or small customer group has a capacity requirement and works closely with the suppliers to provide the desired capacity and system configuration. This structure was adopted for example on PTAT, NPC, Gemini, Southern Cross and Flag Atlantic-1.

Another approach is the private cable venture where a network is put together by a group of venture capital companies or private investors. The VCs are moving into new competitive systems in direct response to market demand, taking over from the traditional planning methods of the past. Examples include Northstar, who have implemented a festoon on the west coast of North America and Global Crossing (AC-1, PC-1, MAC-1, PAC-1, etc.).

with KDD-SCS joining later. This consortium has provided the industry with an essential service for the past 10 years and on 8th March 2000, a new, enhanced, five-year agreement was signed. The five members of this new consortium are Alcatel, Global Marine Systems Ltd, KDD SCS, Pirelli and Tyco Submarine Systems Ltd.

Late-Breaking News

The first submarine cable system simply went from point A to point B. Many new links still do. More recently, systems have begun to evolve into complex integrated networks. The first to follow this approach was APCN, jointly supplied by Alcatel, AT&T SSI, and KDD-SCS, which consists of three interconnected rings. A ring provides instant restoration of the traffic in the event of a link failure by routing data the other way around the loop.

In the 14 years since the laying of UK-Belgium No. 5, the cable industry has moved through a tremendous bit rate explosion. Rocketing from 280Mbit/s in 1986 to 420Mbit/s in 1989, 560Mbit/s in 1990, 2.5Gbit/s in 1994, 5Gbit/s in 1995 to WDM in 1998 and now the truly astronomical capacities of DWDM. We have gone from 1,310nm and 1,550nm re-generating repeaters, to optical amplifiers and until recently this was all driven by demand for transatlantic capacity. However, the Japan-USA system, operating at 640Gbit/s and currently being installed, is probably the first time that the Pacific Ocean has surpassed the Atlantic in terms of leading edge transmission technology. The Atlantic will soon bounce back; Level 3's Project Yellow will be commissioned on 1st September 2000 and will have a design capacity of 1.28 Terabits. This will be followed in 2001 by FLAG Atlantic-1 at 2.4 Terabits. FLAG have also recently announced a new Pacific cable which will go into operation in 2002 and will more than double the existing capacity in the region with a mind boggling 5.2 Terabit capacity. Meanwhile, a 96 fibre pair cable has already been laid between the UK and Belgium, the ultimate capacity of which will only be constrained by investment in terminal transmission equipment and technological development.

Satellite transmission, of course, continues to play an important role. Satellites deliver total coverage of the earth's surface and are ideal for covering broad expanses of low population density and for broadcast applications. Submarine cables, on the other hand, deliver secure communications, very high capacity and constant, irreproachable quality between zones of high density. Inner and outer space do not compete; they are complementary. Having said that, however, in a sense, optical amplification has taken over the world, and by cabling it from top to bottom and from end to end, now it really is a wired world.

The New Future

Fuelled by spiralling demand, especially from the Internet and corporate data traffic, the first four or five years of the new Millennium are destined to be the busiest ever for the fibre optic cable industry, with deployment in each year greater than any year since the advent of submarine cables. Investment between 1999 and 2004 is estimated to total $31.89 billion, nearly twice the total fibre-optic investment as of year-end 1998. The figure for 1999 was $6.9 billion, the industry's most robust year to date, and all indications are that this accelerating trend will continue into 2004.

The amount of fibre deployed in submarine systems with ready-for-service dates in 1999 and 2000 will surpass one million fibre kilometres for each year. These two years will also see the activation of a number of high-fibre-count systems, with some regional systems entering service with as many as 96 fibre pairs.

The capacity explosion is likely to go on unabated with scientists continuing to raid the Greek language to come up with new terminology to describe the extraordinary big bang mathematics of the latest cable technology. Tomorrow's long-haul systems are likely to have eight fibre pairs and beyond with more than 100 wavelengths per fibre pair, each wavelength operating at, at least 10Gbit/s. In the design laboratories, wavelength operating at 40Gbit/s are already a real possibility.

The Optical Era 1986 - 2000

What are we going to do with all those Terabits? No problem at all! An exponential leap in high-speed data, Internet and digital video traffic will soon take care of all that capacity and soon the world's long-distance cable technology will no doubt be upgraded well beyond even one Terabit per fibre pair. There is already talk of updating the Internet's increasingly inadequate plumbing. The world's submarine cable industry will surely be ready, as it always has over the past 150 years, to provide more than adequate pipes and plumbers to meet the demand.

Today's Internet is almost entirely dependant on submarine cables for their speed and international transparency and thanks to the highly reliable and secure networks they provide, the Internet has spawned an entirely new global economy, based on e-commerce. What would Cyrus W Field, John Pender, Daniel Gooch and Charles Tilston Bright think of what we have done with their dream?

We hope they would approve.

Acknowledgements

Compiling a book on such a large topic, even a short one such as this, can never be achieved by one individual. It is the result of a great deal of co-operation on a large scale. The subject is vast and we have only been able to dust the surface of some of the major milestones of the industry. However, a great deal more detailed research of the subject has taken place than can be included in these pages. It is perhaps interesting to note that as our ability to record and store information with the aid of modern technology has increased, our appetite for doing so has diminished. There is a wealth of information on the first 120 years of the industry, not always readily available, but becoming more so as enthusiasts develop web sites. For the true student of the subject, much of the source material used in the book for this early period, can be tracked down and accessed through libraries and museums. However, there is no body of work that covers the last 30 years. This period of the industry, has been the most dramatic and dynamic to date. It is worthy of recording for posterity, and I hope that this book, in some small way, can act as a catalyst to start filling this void.

Firstly, I would like to offer my thanks to Lord Pender for writing the Foreword, and, with Lady Pender, to thank them for their hospitality in allowing me into their home to access their private papers. I am particularly grateful to: my friend and ex-colleague, Nigel Morgan, who wrote the "Birth of the Industry", for his advice, assistance and encyclopaedic knowledge of the subject; to Mary Godwin, Curator of the Porthcurno Telegraph Museum, who wrote the "Telegraph Era", and gave me unfettered access to the Aladdin's Cave that is the Museum archive, during my research; to Mick Green of BT who with Ian Fletcher, in large part, was responsible for writing the "Telephone Era"; and to Sophie Pèpe, Emilie Garin and Sarah Payne from Alcatel's Communications Department in collaboration with their colleagues, for the "Optical Era". I am also indebted to Janet Parsons for compiling the Index and with Gail Clark for their thorough and detailed proof reading of the text. Special thanks must go to Erik Jensen of Great Northern Telegraph for the historical information he provided on his company and for allowing us to use the unique picture of the H C Oersted. In addition, I would like to thank Alan Jeal for some key observations on the technical content of the text and Ann Locker at the Institution of Electrical Engineers' Archive in London for her assistance.

Stewart Ash 2000

Appendix

Industry Chronology 1800 - 2000

1800
Alessandro Volta announces a way of producing a continuous electric current - the "Voltaic Pile".

1820
Christian Oersted announces his observation of the action of an electric current on a magnetic needle. Oersted's observations are explained by André Marie Ampère, who also demonstrates the first coil electromagnet.

1821
Ampère establishes the laws of electromagnetic action. Michael Faraday observes electromagnetic rotation.

1831
Faraday discovers electromagnetic induction (the first transformer), demonstrates conversion of mechanical effort into electricity (magneto-electric induction) and the first electricity generator (copper disc experiment).

1832
Samuel F B Morse realises the possibility of electric telegraph during a dinner conversation with Dr Charles Jackson of Boston, whilst on a voyage from Europe to the United States onboard the sloop *Sully*.

1836
William Fothergill Cooke attends a physics lecture in Heidelberg given by Professor Müncke in which he uses a replica of Schilling's telegraph.

1837
Cooke and Charles Wheatstone agree to form a partnership to exploit the electric telegraph and on 12th June are awarded the world's first telegraph patent. In the USA Morse demonstrates his telegraph system over a short distance.

1838
Cooke and Wheatstone build a 13-mile telegraph system for the Great Western Railway between Paddington and West Drayton.

1840
Wheatstone puts a proposal before a Committee of the House of Commons for a submarine cable from Dover to Calais. The London and Blackwall railway opens, a single line system, it incorporates a Cooke/Wheatstone telegraph system.

1843
Dr William Mongomerie introduces Gutta Percha to the Society for the encouragement of Arts, Manufacture & Commerce in London. Morse spans New York harbour with an electric telegraph and forecasts an Atlantic Cable.

1844
Wheatstone conducts experiments in Swansea Bay with a cable laid between a ship and the lighthouse.

1845
Faraday suggests Gutta Percha as an electrical insulator. Henry Bewley devises a machine to make Gutta Percha tubes, S W Silver devises a method of coating wires with Gutta Percha. On 4th February the Gutta Percha Company is formed. Ezra Cornell lays a 12-mile cable across the Hudson River.

1846
The Gutta Percha Company opens its factory at 18 Wharf Road, City Road Islington in London. Charles West experiments with an India Rubber insulated cable in Portsmouth Harbour.

1848
The Gutta Percha Company receives its first order for a submarine cable, a two-mile length for CV Walker.

1849
On 10th January, CV Walker tests the first Gutta Percha cable in Folkestone Harbour. The English Channel Submarine Telegraph Company is formed.

1850
On 28th August the first submarine cable is laid between Dover and Calais by Jacob and John Watkins Brett.

1851
The Submarine Telegraph Company is formed and lays the first successful Dover to Calais cable. Frederick N Gisborne places a plan for an overland telegraph system before the Newfoundland legislature.

1852
The English & Irish Magnetic Telegraph Company is formed. Gisborne lays a 15-mile cable from Prince Edward Island to New Brunswick.

1853
Charles Tilston Bright lays a cable across the Irish Sea from Port Patrick to Donaghadee. The International Telegraph Company is formed and purchases the *Monarch*, the first vessel to be permanently converted for cable work. Edward & Charles Bright begin nocturnal experiments on their network to simulate an Atlantic cable.

1854
Glass, Elliot & Company is formed at Greenwich. W T Henley establishes a factory in North Woolwich. Bright and Faraday produce papers on retardation and Professor William Thomson takes up this problem. Gisborne goes to New York and meets Cyrus W Field. Field writes to Morse and Maury. Based on their positive responses, he forms the New York, Newfoundland and London Telegraph Company. Field visits England to order a cable for the Cabot Strait and meets John W Brett and Charles Tilston Bright.

1855
Edward Bright surveys the west coast of Ireland and selects Valentia Island to land the Atlantic cable.

1856
The 60-mile Cabot Strait cable is laid by Glass, Elliot & Company. Field, Brett and Bright sign an historic document on 29th September to promote the Atlantic Telegraph Company, which is registered on 20th October. Contracts are awarded

to the Gutta Percha Company, Glass, Elliot and R S Newall to make the cable. The Newfoundland land line between Cape Breton and St John's is completed.

1857

The Atlantic route is re-surveyed by the British Navy confirming Lieutenant Maury's findings including the "Telegraph Plateau". The first attempt by the Niagara to lay the cable fails after 330 miles. The *Agamemnon* and the *Niagara* discharge their cable at Keyham in Plymouth and a further 900 miles is ordered from Glass, Elliot & Company.

1858

The *Agamemnon* and *Niagara* meet in mid Atlantic and after 3 abortive attempts to lay the cable are finally successful. 400 messages are passed before the cable fails. Charles Tilston Bright is knighted at the age of 26. The Red Sea & Telegraph to India Company is formed to link Suez, Aden and Karachi. William Thomson patents his Mirror Galvanometer.

1859

The British Government and the Atlantic Telegraph Company set up a joint Committee to investigate the failure of the Atlantic cable. The Red Sea & Telegraph to India Company fails with losses of £800,000 the British Government had promised the company 4½% on the capital and were obliged to pay interest until July 1908. The total loss was estimated at £1,800,000.

1861

The Government Report on the Atlantic cable is published. Charles Tilston Bright sets up a consultancy company with Josiah Latimer Clark. William Thomson forms a Committee on Electrical Standards. Bright and Clark publish the definitive paper on electrical standards. The British Government installs a cable from Malta to Alexandria.

1863

Siemens & Halske establish a factory in Charlton, South East London.

1864

On 7th April the Telegraph Construction & Maintenance Company is formed with John Pender as Chairman and is awarded a contract for an Atlantic cable. The *Great Eastern* is bought at auction by Daniel Gooch, Thomas Brassey and William Barber. The India Rubber, Gutta Percha and Telegraph Works Company is established in Silvertown. Charles Tilston Bright lays a cable from Fao to Karachi as part of the first England to India Network.

1865

Siemens & Halske at Charlton become Siemens Brothers owned by William and Werner Siemens. William Hooper starts production of rubber insulated core at his Millwall Factory for armouring by R S Newall and W T Henley. The *Great Eastern* fails to lay the Atlantic cable 600 miles short of Newfoundland.

1866

The Anglo-American Telegraph Company is formed. The Telegraph Construction & Maintenance Company make a second Atlantic Cable, which is successfully laid by the *Great Eastern*. She also recovers and completes the 1865 cable.

1867

William Thomson is granted his first patent for the siphon recorder, which provides the first permanent record of weak signals over long oceanic cables.

1868

John Pender stands down as Chairman of the Telegraph Construction & Maintenance Company and is replaced by Daniel Gooch. The Danish-Norwegian-English, the Norwegian-English and the Danish-Russian Telegraph Companies are formed (1). A parliamentary committee is set up to examine the failure of the Red Sea Company. On 31st July Benjamin Disraeli introduces the Telegraph Act, a bill to authorise the Post Office to acquire all telegraph companies in the United Kingdom. The Anglo-Mediterranean Telegraph Company (2) is formed to connect Malta with Alexandria.

1869

Carl Frederik Tietgen merges three companies (1), to form the Great Northern Telegraph Company. The British Indian Submarine Telegraph Company is formed (2). The Falmouth, Gibraltar and Malta Telegraph Company is formed (2). The British India Extension Company is formed to connect Malaya and Singapore (3). The China Submarine Telegraph Company is formed to connect Hong Kong to Singapore (3).

1870

The British Australia Telegraph Company is formed to connect Singapore with Java, Sumatra and Darwin (3). The Marseille, Algiers & Malta Telegraph Company is formed (2). Hooper's Telegraph Works is formed to expand into vulcanised rubber insulated telegraph cables. The Suez-Aden-India cable is laid by the *Great Eastern*. The Madras-Penang-Singapore cable is laid. The Porthcurno cable station is opened. Tietgen forms the Great Northern China and Japan Extension Telegraph Company; it lands the first submarine cables in Hong Kong and China. On 23rd June the first telegram is sent from London to Bombay over an all-submarine cable route. The British Government takes control of all telegraph companies operating in the United Kingdom.

1871

Great Northern land the first submarine cable in Japan at Shembon.

1872

On 1st January Great Northern open a telegraph service between Europe and the Far East via the trans-Siberia telegraph system. Great Northern launch the *H C Oersted* the first purpose built cable repair ship. On June 1st John Pender merges his four companies (2), to form the Eastern Telegraph Company. On 21st October the London to Adelaide telegraph goes into service.

1873

The *Hooper* is launched; the first purpose built cable laying vessel. John Pender merges three of his companies (3) to form the Eastern Extension, Australasia and

Appendix

China Telegraph Company. The Brazilian Submarine Telegraph Company is formed to connect Portugal to Brazil via Madeira and the Cape Verde Islands. Charles Tilston Bright completes the Eastern Caribbean network.

1874
The first Japanese manufactured submarine cable is laid. Siemens Brothers launch the cableship *Faraday*; it is the first twin screwed cableship and the first to use taut wire.

1875
The Central American Telegraph Company is formed to lay a cable from Brazil to Demerara (British Guiana). This is the first system to have a branching unit, which had a spur to Cayenne in French Guiana. The first submarine cable is laid between Australia and New Zealand. Alexander Muirhead invents a method of Duplex working. An International Telegraph Conference is held in St Petersburg to agree a set of principles for telegraph standards.

1876
On 3rd March Alexander Graham Bell receives a patent for the telephone.

1877
The New South Wales and New Zealand Governments begin plans to build a Pacific Cable. The 1865 Atlantic cable fails and is abandoned.

1879
The Eastern and South African Telegraph Company is formed. Brass tape is introduced into cable design as a protection against teredo worms.

1883
Great Northern and the Eastern Telegraph Company enter a joint purse agreement for Europe to Far East traffic.

1885
Oliver Heaviside demonstrates that, for voice transmission, cable capacitance can be countered by loading inductance evenly along the length of the cable.

1887
Heinrich Hertz confirms the existence of electromagnetic (radio) waves.

1889
The *Great Eastern* is sold to Henry Bath & Sons to be broken up, a task which would take two years to complete.

1890
The British Post Office takes over the Submarine Telegraph Company.

1891
The London to Paris telephone service goes into operation over the world's first submarine telephone cable. The Société Générale des Téléphones opens its submarine cable factory at Calais to armour core made at Bezons.

1892
The British Admiralty survey the route for the Pacific cable.

1893
The Société Générale des Téléphones becomes the Société Industrielle des Téléphones and their ship the François Argo lays the Queensland to New Caledonia cable.

1895
Frederick Creed invents the Morse keyboard perforator, the forerunner of the teleprinter.

1896
Guglielmo Marconi receives the world's first patent for wireless telegraphy.

1899
Marconi sends wireless telegraph signals across the Channel. The Pacific Cable Committee issues its report endorsing the building of a Government owned cable.

1900
Marconi takes out his famous Four Sevens patent, which solved the problem of tuned circuits.

1901
The Eastern Telegraph Company opens a service from Durban to Perth via Mauritius. Marconi demonstrates wireless telegraph across the Atlantic from Poldhu in Cornwall to Signal Hill in Newfoundland.

1902
The Telegraph Construction & Maintenance Company launch the *Colonia* to lay the Pacific cable. The *Colonia* lays the Pacific cable including the Bamfield, British Colombia to Fanning Island link which, at 3,458 nautical miles was the longest submarine telegraph cable ever laid. On 31st October the first message is sent from Ottawa to New Zealand and back again. *HMTS Iris* is launched to maintain the cable.

1904
Dr J Ambrose Fleming, Professor of Electrical Engineering at University College, London, patents the thermionic valve diode.

1906
Lee de Forest develops the thermionic triode valve.

1907
Marconi opens the first wireless telegraph service across the Atlantic, from Clifden in Ireland to Glace Bay in Canada.

1910
Siemens Brothers manufacture and lay the first loaded telephone cable from Dover to Calais using Pupin Coils. Designed by Dr Michael Pupin.

1912
The Telegraph Construction & Maintenance Company lay a four-core continuously loaded telephone cable from England to France.

1917
C S Franklin of Marconi's achieves 20 miles transmission of beam radio in Carnarvon.

1919
Marconi proposes an Imperial Wireless network to the British Government, which sets up a Committee to consider it. C S Franklin achieves 93-mile beam radio transmission from Hendon to Birmingham.

1921
The Western Electric Company patent Permalloy, a nickel-iron alloy for loading cables. A 54 valve beam radio is tested by Marconi's between Canarvon and Australia.

1923
On 14th January the first, one way, radio-telephone call is made between Rocky Point, New York and the STC factory in New Southgate. The Telegraph Construction & Maintenance Company patent Mumetal, a copper-nickel-iron alloy for loading cables.

1924
A Mumetal continuously loaded telegraph cable is installed between New York and the Azores. It operates at 1,500 words a minute, the fastest ever achieved across the Atlantic.

1927
The British Post Office, establish a beam wireless telegraph service to Australia, Canada, India and South Africa. The first radio telephone service across the Atlantic goes into operation between Rugby and Rocky Point, New York.

1929
On 30th September British international cable and wireless interests are combined with the forming of Imperial & International Communications Ltd.

1933
Imperial Chemical Industries (ICI) discovers Polyethylene.

1934
Imperial & International Communications Ltd is renamed Cable & Wireless Ltd.

1935
Submarine Cables Ltd is formed by merging the Telegraph Construction & Maintenance Company with the submarine cable division of Siemens Brothers.

1938
Alec Reeves patents Pulse Code Modulation. The Société Industrielle des Téléphones cable factory at Calais comes under the control of Câbles de Lyon.

1943
On 24th June HMTS Iris (II) recovers the Anglesey to Port Erin coaxial cable and inserts the first telephone repeater.

1947
Cable & Wireless is nationalised. Submarine Cables Ltd lay a 1.7-inch air cored coaxial cable between England and Holland. Bell Laboratories invent the transistor.

1948
The Commonwealth Telegraphs Agreement is implemented in which the Governments of Australia, Canada, New Zealand, South Africa, India and Southern Rhodesia take control of their telecommunications assets.

1950
To celebrate the Centenary of the Submarine Cable Industry an exhibition is held at the Science Museum in London, where a telegraph signal was sent 33,871 miles, relayed five times and returned to London in 53.6 seconds.

1953
Simplex Wire and Cable Company open their submarine cable factory in Newington, New Hampshire.

1954
Submarine Cables Ltd establish a cable factory at Erith.

1956
STC opens its first submarine cable factory in Southampton Docks. The installation of the first transatlantic telephone system is completed at 20.52 GMT on 14th August. It had a capacity of 36 x 4kHz.

1957
Russia launch Sputnik 1.

1960
Repeater parachutes are successfully tested in Loch Fyne by the Ariel.

1961
CANTAT is installed between the UK and Canada, the first system to use lightweight cable. It had an initial capacity of 80 x 4kHz.

1962
Active satellites Telstar and Relay with amplification and re-transmission capability are launched. STC open a second cable factory in Southampton Docks to make American designed cables. 3kHz channels are introduced on TAT-1, TAT-2 and CANTAT.

1963
The Communications Satellite Act in the USA sets up COMSAT. CS Long Lines is launched, she is equipped with a caterpillar Linear Cable Engine. The first geosynchronus orbit satellite Syncom II is launched. The first transpacific telephone cable COMPAC is installed.

1965
The Early Bird Telecommunications satellite is launched.

1966
Dr Charles Kao and Dr George Hockham publish their famous paper on fibre optic transmission.

1967
SEACOM goes into operation.

1968
SAT 1 is laid from Cape Town to Lisbon. This is the last major system to use thermionic valve repeaters.

Appendix

1970
STC purchase Submarine Cables Ltd and form their Submarine Cable Division. Alcatel Submarcom is formed. Corning announce mono mode fibres with losses of 1db/km. Panish and Hayashi of Bell Labs demonstrate the first semi-conductor laser to operate continuously at room temperature.

1971
The British Post Office install a 14-wheel pair Linear Cable Engine on CS *Alert*.

1973
Cable & Wireless install an 18-wheel pair linear engine on CS *Mercury* to install CANTAT II (1,840 x 3 kHz).

1975
Corning produce 4km lengths of fibre with losses of 4db/km. Bell Labs set up the world's first field trial of optical telecommunications. A 2,100ft cable containing 144 fibres is set up between two offices in Atlanta Georgia, operating at 820nm and 45Mbit/s.

1977
STL install the first commercial fibre optic telecommunications system for the British Post Office. A 4km length between Hitchin and Stevenage, which operated at 850nm and 140Mbit/s. Pencan 3 is installed by STC between Grand Canaria and mainland Spain. It is the longest 45MHz system to be laid.

1980
STC install the world's first experimental fibre optic submarine system in Loch Fyne.

1983
Cable & Wireless (Marine) Ltd is formed as a wholly owned subsidiary. TAT-7 is installed, the last transatlantic coaxial telephone system (4,200 x 3Khz).

1984
The Federal Communications Commission in the USA breaks up AT&T.

1985
Drs D Payne and S B Poole develop the Erbium Doped Optical Amplifier at Southampton University. SEA-ME-WE is installed. It is the longest submarine telephone system in the world.

1986
AT&T deploy their first experimental submarine optical system Optican. Initial trials are completed on the SMD cable burial plough. The first international optical fibre submarine system is installed, UK-Belgium No. 5, operating at 1,310nm and 280Mbit/s. Alcatel install their first optical system, Mainland to Corsica.

1987
BT Marine Ltd is formed as a wholly owned subsidiary.

1988
The first transatlantic fibre optic system TAT-8 is installed, operating at 1,310nm and 280Mbit/s. The repeaterless system UK Channel Islands No.7 is installed. It operates over 133km at 1,550nm and 140Mbit/s.

1989
PTAT-1 the first private Atlantic optical cable goes into operation operating at 1,310nm and 420Mbit/s. STC establish a cable factory in Portland Oregon. The Universal Jointing Consortium is formed.

1990
TAT-9 is installed. It is the first repeatered system to operate at 1,550nm and 560Mbit/s, it also uses submerged multiplex equipment for the first time. NPC the first private cable across the Pacific is installed from Japan to the USA, operating at 420Mbit/s.

1991
Alcatel Cables is formed absorbing Cables de Lyon.

1992
Alcatel establish a cable factory in Port Botany, New South Wales. Tasman 2 is installed from Australia to New Zealand.

1994
The repeaterless system CELTIC is installed, it operates over 270km at 2.5Gbit/s. Alcatel purchase STC Submarine Systems to form Alcatel Submarine Networks. SEA-ME-WE 2 is installed. CANTAT-3 goes into operation. It is the highest capacity regenerating repeater system operating at 2.5Gbit/s on 3 fibre pairs.

1995
TAT-12 is installed, it is the first transatlantic system to use optical amplifiers. The Rioja system is installed, an optically amplified system operating at 2.5Gbit/s, it is the first system to deploy a

1996
ROPA in a repeaterless span. Cable & Wireless (Marine) Ltd and BT Marine Ltd merge to form Cable & Wireless Marine Ltd.

On 2nd July Alcatel close their cable factories in Southampton Docks, the last factories making submarine cable in the United Kingdom. TAT-13 is installed completing the first self-restoring Atlantic ring. Tyco Submarine Systems Ltd purchase AT&T SSI.

1998
Forward Error Correction and Wave Division Multiplex is introduced to upgrade the capacity of TAT-12/13. SEA-ME-WE 3 is installed. It is the first system to be initially equipped with WDM technology, operating at 2.5Gbit/s and 4 wave-lengths per fibre pair.

1999
Global Crossing Ltd purchase Cable & Wireless Marine Ltd and rename it Global Marine Systems Ltd. Tyco Submarine Systems Ltd purchase Temasa.

2000
FT Marine is formed as a wholly owned subsidiary of France Telecom. A new Universal Consortium Agreement is signed. Japan-US is installed in the Pacific. TAT-14 is installed in the Atlantic, followed by Project Yellow, with a design capacity of 1.24 Tbit/s it is the highest currently in operation.

The 28th August is the 150th Anniversary of the first international submarine cable lay.

Bibliography

The Private Papers of Lord Pender of Porthcurnow

The Archive of the Cable & Wireless Porthcurno and Collections Trust

The Archive of the Institution of Electrical Engineers in London

The Electrical Telegraph Popularised, Dr Dionysius Lardner, Walter & Maberly, London 1855

The Story of the Telegraph and a History of the Great Atlantic Cable, Charles F Briggs and Augustus Maverick, Rudd and Carleton, New York 1858

Enquiry into Submarine Cables - Report of the Joint Committee appointed by the Lords of the Privy Council for Trade and the Atlantic Telegraph Company, HMG Privy Council and the Atlantic Cable Company, HMSO, London 1861

The Atlantic Telegraph, W H Russell, Day & Son Ltd., London 1865

A History of Electrical Telegraphy to the Year 1837, J J Fahie, E&F N Spon, London 1884

Cyrus W Field, His Life and Work, Isabella Field Judson, New York 1896

Submarine Telegraphs, Their History, Construction and Working, Charles Bright, London 1898

The Life Story of Charles Tilston Bright, Civil Engineer, Charles Bright, Archibald Constable & Co Ltd. 1908

Submarine Cable Laying and Repairing, HD Wilkinson, 'The Electrician' Printing and Publishing Co, London 1910

One Hundred Years of Submarine Cables, G R M Garratt, His Majesty's Stationery Office 1950

The Telcon Story, G L Lawford and L R Nicholson, The Fanfare Press Ltd. 1950

The Great Iron Ship, James Dugan, Harper & Bros, New York 1953

Siemens Brothers, 1858-1958 - An Essay in the History of Industry, J D Scott, Weidenfield & Nicholson, London 1958

Cooke and Wheatstone and the Invention of the Electric Telegraph, Geoffrey Hubbard, Routledge & Kegan Paul, London 1965

Michael Faraday, A Biography, L. Pearce Williams, Chapman & Hall, London 1965

From Semaphore to Satellite, ITU, Geneva 1965

A Century of Service, Cable and Wireless 1868-1968, K C Baglehole, Cable & Wireless, London 1969

Isambard Kingdom Brunel, L T C Rolt, Penguin/Pelican, Harmonsworth 1970

Sir Daniel Gooch: Memoirs and Diary, Roger Burdett Wilson, David & Charles, Newton Abbott 1972

Submarine Telegraphy, The Grand Victorian Technology, Bernard S Finn, Her Majesty's Stationary Office 1973

The Electric Telegraph - A Social and Economic History, Jeffrey Kieve, David & Charles, Newton Abbott 1973

Sir Charles Wheatstone, Brian Bowers, HMSO and Science Museum, London 1975

The Old Telegraphs, Geoffrey Wilson, Phillimore, Chichester 1976

Cableships and Submarine Cables, K R Haigh, STC Submarine Systems 1978

Girdle Round the Earth, Hugh Barty-King, William Heinman Ltd., London 1979

The Power of Speech, A History of Standard Telephones and Cables 1883-1983, Peter Young George Allen & Unwin, London 1983

The History of Electric Wires and Cables, R M Black, Peter Peregrinus, 1983

Energy and Empire - A Biographical Study of Lord Kelvin, Crosbie Smith and M Norton Wise, Cambridge University Press 1989

1990 World's Submarine Telephone Cable Systems, Herbert H Schenck & Dr Leo Waldwick, US Department of Commerce 1991

Cent Ans de Câbles, Stéphane & Elisabeth Curveiller, Alcatel 1991

The Life of Captain Robert Halpin, Jim Rees, Dee-Jay Publications Arklow, County Wicklow 1992

BT Marine and Before, Roger Hardingham, BT (Marine) Limited 1993

Bibliography

From Dots and Dashes to Tele and Datacommunications, Great Northern Telegraph, Copenhagen, Denmark 1994

The Communication Miracle, John Bray, Plenum, London 1995

Optical Line Systems, Douglas Maclean, Wiley, London 1996

Un Siècle d'histoire du Câble à Lyon, Gérard Hauser, France 1997

Semaphores to Short Waves, Frank A J L James, RSA, London 1998

Atlantic Sentinel, D R Tarrant, Flanker Press Ltd. St. John's Newfoundland 1999

Papers and Journals

Transatlantic Telephone Cable, supplement to Post Office Telecoms Journal, Autumn 1956

The Anglo-Canadian Transatlantic Telephone Cable (CANTAT), R Halsey, POEJJ, April 1963

The Anglo-Canadian Transatlantic Telephone Cable (CANTAT): Cable Laying Gear, E Clark & K Chapman, POEJJ, April 1963

The Anglo-Canadian Transatlantic Telephone Cable (CANTAT): Cable Development, Design and Manufacture, R Brockbank, E Clark & F Jones, POEJJ, April 1963

The Communication Miracle, Proposed Commonwealth Pacific Telephone Cable System, POEJJ, Vol. 53 April 1960 – January 1961

Optical Communication: Dialetric – Fibre Surface Waveguides for Optical Frequencies, K C Kao & G A Hockham, PROC. IEE. Vol 113 No 7, July 1966

Manufacturing Technology of optical fibres, C J Swan, published in GLASS, August 1982

Optical Fibre Communications So Far, Robert L Williamson, published in Interdisciplinary Science Reviews Vol. 8, no 1983, Wiley Heyden Ltd 1983

1300nm 140Mbit/s Monomode fibre optics in practice, P J Howard, published in Communications International, January 1984

Co-axial Submarine Telecommunications Systems, STC, July 1985

Submarine Cable Design Requirements and Testing, A Gould, published in British Telecommunications Engineering Vol. 5, July 1986

The Loch Fyne Optical-Fibre Cable System, R E Ferris and R D Channon, British Telecommunications Engineering Vol. 5, July 1986

Protection and Installation Techniques for Buried Submarine Cables, C Cole and R Struzyna, British Telecommunications Engineering Vol. 5, July 1986

Dispersion Shifted Fibre for Long Unrepeatered Submarine Systems, published in Electrical Communications Vol. 61 No. 4, 1987

UK-Belgium No 5, Part 1 Marine Aspects of the Route Selection, R Whittington, British Telecommunications Engineering Vol. 5, July 1986

UK-Belgium No 5, Part 2 Technical and Installation Aspects, R D Channon, British Telecommunications Engineering Vol. 5, July 1986

TAT-8 An Overview, R L Smith and R Whittington, British Telecommunications Engineering Vol. 5, July 1986

Dix ans de Câbles Sous-Marins à fibre optique (Première Partie), Alain Paul Leclerc, Les Amis des Câbles Sous-Marins, Bulletin 9, July 1998

Dix ans de Câbles Sous-Marins à fibre optique (Deuxième Partie), Alain Paul Leclerc, Les Amis des Câbles Sous-Marins, Bulletin 13, August 1999

Index

Aberdeen, 63
AC-1, 83
Admiralty, 13, 41
Africa, 38
Agamemnon, 28, 30, 31
Alcatel, 7, 66, 71, 72, 73, 74, 75, 76,
Alert, 21, 64
Alexandria, 37, 61
Algiers, 61
America, 7, 16, 24, 31, 38, 47, 59, 73, 75, 83
American Civil War, 31
American Telephone & Telegraph Company, 50
Ampere, Andre-Marie, 10
Anglo-American Telegraph Company, 33
ANZCAN, 58
APCN, 83
Apollo, 29
AT&T, 50, 53, 55, 58, 59, 60, 64, 65, 66, 71, 72, 73, 77, 78, 80, 82, 83
Atlanta, 67
Atlantic, 7, 16, 21, 23, 24, 25, 27, 29, 30, 31, 33, 34, 37, 39, 43, 50, 58, 59, 72, 75, 84
Atlantic Telegraph, 7, 27, 29, 33
Auckland, 57, 58
Australia, 7, 37, 40, 41, 44
Azores, 34
Baby Bells, 73
Bamfield, 41, 43
Beales, Dr Keith, 67
Belgium, 54, 75, 84
Bell, Alexander Graham, 49
Bell Telephone, 47, 51
Bergen, 63
Berlin, 15
Bermuda, 7, 65
Bewley, Henry, 15
Bezons, 40, 61
Bidder, George Parker, 13
Birkenhead, 30
Birth of the Industry, 7, 9
Blazer, 19
Blue Ribbon, 72
Bombay, 36, 37
Botany Bay, 74
Boulogne, 19
Bourseul, Charles, 49

Brett, Brothers, 7, 17, 19, 21
Brett, Jacob, 7, 16
Brett, John Watkins, 7, 16, 17, 19, 25, 27, 29
Bright, Charles, 55
Bright, Charles Tilston, 21, 22, 23, 25, 27, 29, 31, 37, 55, 85
Bright, Edward, 21, 22, 23, 29
Brisbane, 41
Britain, 6, 17, 25, 29, 31, 34, 38, 40, 45, 47, 66, 71
British Admiralty, 41
British Association for the Advancement of Science, 23
British Empire, 6, 37
British Indian Submarine Telegraph Company, 33
British Telecom, 7, 65, 71, 77, 78, 80, 83
Brunel, Isambard Kingdom, 3, 33
Brussels, 13
BT Marine, 80, 82
Buchanan, President James, 31
Cable & Wireless, 7, 22, 44, 45, 54, 58, 64, 65, 66, 73, 80
Cable Venture, 56, 59
Cairns, 58
Calais, 17, 40, 47, 61
Camden Town, 13
Canada, 16, 41, 44, 51, 54, 55, 57, 58, 75
Canadian Overseas Telecommunications Corporation, 50, 54
Canary Islands, 59, 77
Cannes, 61
Canning, Samuel, 34
CANTAT, 54, 55, 57, 58
CANTAT-2, 59
Canterbury, 41
Cap Gris-Nez, 13, 17
Cape Breton Island, 25
Cape Ray, 24, 25
Carmichael, Sir James, 19
CASCADE™, 80
Castor, 78
Celtic, 75
Chamberlain, Joseph, 41
Charlton, 33
China, 39, 40

China-USA, 83
Clarenville, 51, 52, 53, 54
Clifden, 44
Colombo, 61
Colonia, 41
Commonwealth Telecommunications Conference, 57
Commonwealth Trade and Economic Conference, 57
COMPAC, 57, 58
Compagnie Industrielle des Téléphones, 61
Cooke, William Fothergill, 10, 11, 13, 14, 17, 21
Corfu, 33
Cornell, Ezra, 14
Corning, 66, 67
Cornwall, 33, 37, 43
Corsica, 72
County Kerry, 31
Crampton, Thomas Russell, 7, 19, 21
Cuckmere, 49
Dagenham, 52
Daniell, Professor, 9
Danish-Norwegian-English Telegraph Company, 33, 39
Danish-Russian Telegraph Company, 39
Darwin, 41
Davey, Edward, 10, 13
Davey, Humphrey, 9
De Magnete, 9
Denmark, 39, 54, 74, 75
Dickens, Charles, 6
Dieppe, 49
Djibouti, 61
Dollis Hill, 53, 55
Dominia, 43
Donaghadee, 23, 33
Doubtless Bay, 41
Dover, 13, 17
Dowry, Boulton Paul, 64
Dublin, 14, 23
Dunlop & Co, 38
Early Bird, 58
East India Company, 14
Eastern and Associated Telegraph Companies, 36, 38, 40, 41, 44, 46
Eastern Extension Company, 40, 41

Eastern Telegraph Company, 37, 38, 41, 43
Edwards, Francis, 17
Electric, 13, 14, 21, 22
Electric Telegraph Company, 13, 21
Elektron, 9
Eletrra, 44
EMOS, 74
England, 6, 7, 14, 16, 17, 19, 22, 23, 24, 25, 27, 36, 37, 39, 43, 53
English and Irish Magnetic Telegraph, 7, 22
English Channel Submarine Telegraph Company, 17
Erbium Doped Fibre Amplifier, 61
Ericsson, 74
Erith, 52
ESPADON, 80
Europe, 21, 37, 44, 52, 69, 73, 74, 75
Euston, 13
Faktor, Marc, 67
Fanning Island, 41, 43
Faraday, 39, 65
Faraday Hills, 39
Faraday, Michael, 9, 11, 15
Father Eastern, 38
Fayal, 37
Federal Communications Commission, 73
Field, Cyrus W, 24, 25, 27, 29, 31, 34, 85
Fiji, 41, 57, 58, 65
Finn, Bernard S, 29
FLAG Atlantic-1, 83, 84
Fleming, Sandford, 41
Folkestone, 16
Foochow, 40
Fort Lee, 14
Fox, Charles, 17
France, 6, 7, 14, 16, 17, 19, 23, 25, 40, 41, 49, 54, 58, 59, 69, 70, 71, 72, 73, 74, 75
France Telecom, 65, 66, 72, 78, 80, 83
France-UK No. 4, 78
Francois Arago, 40
Franklin, Benjamin, 9
Fujitsu, 66
Gateshead, 19
Gemini, 82, 83
General Oceanic & Subterranean Electric Printing Telephone Company, 16

Index

Georgia, 67
Germany, 15, 47, 54, 70, 74
Gilbert, William, 9
Gisborne, Frederick, 24, 25
Glace Bay, 44
Glasgow, 29, 31, 38, 71
Glasgow University, 30
Glass, Richard Atwood, 34
Glass, Elliot, 21, 25, 30, 33
Global Crossing, 7, 80, 83
Global Marine Systems Ltd, 7, 80, 83
Goliath, 17
Gooch, Daniel, 33, 34, 37, 47, 85
Gray, Stephen, 9
Great Eastern, 32, 33, 38
Great Exhibition, 19
Great Northern China and Japan Extension Telegraph Company, 39
Great Northern Telegraph Company, 33, 39, 40
Great Western Railway, 13
Greece, 33, 74
Greenwich, 7, 14, 25, 28, 30, 33
Guam, 58
Gulf of St. Lawrence, 25
Gutta Percha, 7, 13, 14, 15, 16, 17, 19, 21, 30, 33, 34, 38, 39, 49
H C Oersted, 39, 40
Halifax, 24
Halle, 10
Hancock, Charles, 14, 15, 19
Harlow, 67, 69
Harmonic Telegraph, 49
Harrow, 21
Havana, 51
Hawaii, 65
Heart's Content, 33
Heaton, John Henniker, 41
Heaviside, Oliver, 49
Heidelberg University, 10
Henley, W T, 33, 39
Heurtley, E S , 36
Hillside, 53
Hitchin, 70
Hockham, Dr George Combine, 66, 69, 70
Holland, 54
Holyhead, 50

Hong Kong, 39, 40, 58
Honolulu, 57, 58
Hooper, 38
House, Prof. Royal E, 16, 17, 19, 23
Hudson, Capt W L, 30
Hyde Park, 19
Hydrogen, 81, 82
ICI, 47, 49
Imperial and International Communications, 44
Imperial Pacific Cable Committee, 41
Imperial Wireless and Cable, 44
India, 31, 37, 44, 61
India Rubber, Gutta Percha and Telegraph Works Company, 33, 38
Industrial Revolution, 6
INMARSAT, 65
Innovator, 79
Internet, 6, 70, 76, 84, 85
Ireland, 23, 24, 27, 31, 33, 44
Iris, 41
Iris (II), 50
Iris (III), 71
Isle of Man, 50
Isles of Scilly, 33
Israel, 74
Italy, 74
Japan, 39, 47, 58, 69, 70, 71, 74, 75
Japan Inc, 66
Japan-USA, 83, 84
Jeddah, 61
Jenkin, Fleeming, 34
Kao, Dr Charles, 66, 69, 70
Karachi, 37
KDD, 71, 83
Kelvin, Lord, 10
Kent, 52
Key West, 51
King's College, 9, 11, 13
Kokusai Denshin Denwa, 66
Korea, 75
Kremlin, 40
Küper, 7, 25
L'Illustration, 49
Lady Carmichael, 21
Leech, Capt. W H, 53
Lenin, Vladimir Ilich, 40

Les Câbles de Lyon, 61
Leyden Jar, 9
Lime Street, 11
Liverpool, 11, 22, 23, 29
Liverpool and Manchester Railway, 11
Loch Fyne, 65, 71, 72
London, 9, 11, 13, 15, 16, 21, 23, 29, 30, 33, 37, 41, 52, 53, 54, 66, 70
London and Blackwall Railway, 13
London and North-Western Railway, 21
Long Lines, 64
MAC-1, 83
Madras, 41
Madang, 58
Magnetic, 7, 22, 23, 25, 27, 30
Mainland, 72
Malaya, 14
Malaysia, 58
Malta, 33, 37
Manchester, 11, 22, 23, 29
Marconi, Guglielmo, 42, 43, 44
Marseille, 61
Mauley, Lord de, 19
Maury, Lt Matthew F, 24
Mayor of Liverpool, 29
Medan, 61
Mediterranean, 25, 33
Mercury, 58, 64
Monarch (II), 38
Monarch (IV), 53, 54, 58
Montgomerie, Dr William, 14
Montreal, 54, 57
Morse Code, 43
Morse, Samuel F B, 14, 24, 27
Muirhead, Alexander, 37
Müncke, Professor, 10, 11
Munich, 10
Nagasaki, 39, 40
Napoleon, Emperor III, 17
National Observatory, 24
NEC, 58, 66, 74
New Brunswick, 24
New Caledonia, 40
New Hampshire, 47, 52
New Jersey, 53
New South Wales, 40

New Southgate, 44
New York, 14, 24, 25, 29, 44, 51, 69
New York, Newfoundland and London Telegraph Company, 24, 27
New Zealand, 41, 74
Newall, 19, 22, 30, 33, 39
Newfoundland, 23, 24, 25, 27, 31, 33, 34, 43, 51, 52, 54, 61
Newington, 47
Newns, Dr George, 67
Niagara, 30, 31
Nice, 61
NKT, 74
Norddeutsche, 47
Norfolk Island, 41, 58
North Sea, 78
North Star, 83
North Woolwich, 33, 53
Northern Telecom, 75
Norway, 54, 74
Norwegian-English Telegraph Company, 39
Nova Scotia, 16, 51, 54
NPC, 74, 83
NTT, 71
Oban, 52, 53, 54
Ocean Cable Company, 47, 66
Ocean Layer, 61, 64
Oersted, Christian, 10
Optical Era, 7, 69, 74, 80
Optican, 72, 77
Oregon, 75
PAC-1, 83
Pacific, 40, 41, 43, 57, 58, 84
Pacific Cable Board, 41, 43, 44
Paddington, 13
Palermo, 61
Paris, 13, 29
Payne, Dr D, 75
PC-1, 83
Peabody, George, 29
Penang, 41
PENCAN 3, 59
Pender, Sir John, 7, 22, 30, 33, 34, 36, 37, 38, 39, 40, 45, 47, 85
Penguin Island, 54
Penmarch, 61

Persian Gulf, 37
Philadelphia, 9
Pickering, Charles, 29
Pirelli, 74, 83
Poldhu, 43
Poole, Dr S B, 75
Port Glasgow, 38
Porthcurno, 35, 37, 43
Portland, 51, 75
Portpatrick, 23, 33
Portsmouth, 13, 14
Post Office, 13, 21, 44, 47, 49, 50, 51, 53, 55, 63, 64, 66, 67
Prince Edward Island, 24
Princess Clementine, 16
Prussia, 15
PTAT-1, 74, 75, 83
Queen Elizabeth, 9
Queen Victoria, 31, 34
Queensland, 41
Red Rover, 19
Red Sea and India Telegraph Company, 33
Reiss, Philip, 49
Reeves, Alec, 61
Ricardo, John Lewis, 13
Rioja, 75, 82
River Hudson, 14
River Neva, 10
Rockall Bank, 53, 54
Rocky Point, 44
Ronalds, Francis, 9
ROV, 78
Royal Institution, 9
Royal Society, 15
Russia, 39, 75
Russian Embassy, 10
Russian Revolution, 40
S W Silver & Co, 14
Sabah, 58
San Francisco, 58
Sardinia, 33
Scandinavia, 39
Schilling, Baron Pawel Lwowitsch Schilling von Canstatt, 10, 11,
Schroder, Messrs., 29
Schweigger, Johann, 10
Scotland, 33, 71
Scots' Magazine, 9
SEACOM, 58

SEA-ME-WE, 61
SEA-ME-WE-2, 75
SEA-ME-WE-3, 83
Second World War, 44, 45, 47
Shanghai, 39, 40
Siemens, 15, 19, 33, 34, 39, 49, 74
Siemens, William, 15, 65
Silvertown, 33
Silvertown, 38
SIMEC, 78
Simplex, 47, 52, 53, 66
Singapore, 29, 58, 61, 65
Slough, 13
Société Générale des Téléphones, 40
Société Industrielle des Téléphones, 40
Society of Arts, 14
Soil Machine Dynamics, 78
South Eastern Railway Company, 15
South Foreland, 19
Southampton, 55, 65
Southampton University, 75
Southern Cross, 83
Southern United Telephone Cables, 52
Southport, 41
Spain, 58, 59, 65, 75, 80
St John's, 24, 25
St Petersburg, 10
Standard Telephones and Cables, 53
STC, 44, 53, 55, 58, 59, 60, 61, 66, 69, 71, 73, 74, 75
Stevenage, 70
STK, 74
STL, 69, 70
Strearns, J B , 37
Submarine Cables Ltd, 34, 52, 55, 56
Submarine Telegraph Company, 7, 17, 19, 21, 25
Submarine Telegraphy, The Grand Victorian Technology, 29
Suez, 31, 33, 61
Suva, 41, 43
Swansea Bay, 14
Sweden, 54, 74
Switzerland, 54
Sydney, 57, 58
Sydney Mines, 51, 52, 53, 63
TASMAN-2, 74
TAT-1, 7, 50, 55, 57, 61, 63, 75
TAT-2, 55, 61
TAT-3, 58

TAT-4, 58, 61
TAT-5, 58
TAT-6, 59, 61, 66
TAT-7, 59, 60, 61
TAT-8, 72, 73, 75, 78, 82
TAT-9, 74, 75, 82
TAT-12/13, 75, 76, 80, 83
TAT-14, 83
Tawell, John, 13
Taylor, H A, 37
TC&M, 21
Telcon, 21, 33, 34, 37, 41, 43, 47, 66
Telegraph Construction & Maintenance Company, 7, 21, 23
Telegraph Era, 7, 29, 47, 66
Telegraph Plateau, 24
Telephone Era, 7, 49, 66, 77
Temasa, 80
Terrenceville, 63
Thames, 17, 33, 52
Thatcher, Margaret, 73
The Netherlands, 75
The Pacific Cable Act, 43
The Times, 17
Thomson, William, 10, 30, 31, 34, 36
Tietgen, Carl Frederik, 39, 40
Tokyo, 69
TPC-3, 74
TPC-4, 75
TPC-5, 82, 83
Trencher, 78
Trinity Bay, 31
TSSL, 80
Turkey, 37, 74
Tweeddale, Marquis of, 41
Tyco Submarine Systems, 80, 83
UK-Belgium No. 5, 72, 78, 84
UK-Belgium No. 6, 78
UK-Channel Islands No. 7, 72
UK-Denmark No. 4, 78
United Arab Emirates, 61
Universal Jointing Consortium, 82
University of Copenhagen, 10
University of Glasgow, 31
USA, 44, 47, 51, 52, 54, 58, 59, 66, 69, 70, 71, 72, 73, 74, 75
Valentia, 23, 31, 33
Vancouver, 41, 57
Vercors, 61

Victorian, 15, 29
Vigo, 65
Vladivostok, 39, 40
Volta, Alessandro, 9
Voltaic Pile, 9, 10
Walker, C V, 15, 16
Wapping, 19, 20
West, Charles, 14
West Drayton, 13
West Indies, 38
Western Electric, 34, 53, 66
Western Union Telegraph Company, 34, 47
Westmeath, 40
Wheatstone, Charles, 10, 11, 13, 14, 15, 17, 23
Whitehouse, Dr. Edward, 31
Widgeon, 17
Wilkins and Weatherley, 19, 20
Wollaston, Charlton James, 17
Yarmouth & Norwich Railway, 13
Yokohama, 47